Winners and Losers

Winners and Losers:
Home Ownership in Modern Britain

Chris Hamnett
King's College London

LONDON AND NEW YORK

First published in 1999 by UCL Press

Reprinted 2003 by Routledge
11 New Fetter Lane
London, EC4P 4EE

Routledge is an imprint of the
Taylor & Francis Group

British Library Cataloguing-in-Publication Data
A catalogue record for this book is available from the British Library.

Library of Congress Cataloguing-in-Publication Data are available.

ISBNs:
1-85728-333-3 HB
1-85728-334-1 PB

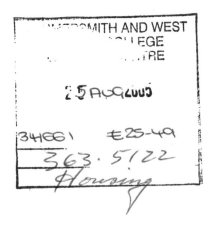
Typeset by Graphicraft Ltd, Hong Kong.
Printed by Antony Rowe Ltd, Eastbourne.

Contents

List of Figures   ix
List of Tables   xi
Acknowledgements   xiii
List of Abbreviations   xiv

1 Home Ownership in Britain: Riding the Roller
   Coaster   1
   The Booms of the 1970s and 1980s   1
   The 1990s Slump   5
   The Recovery   8
   Key Arguments of the Book   12

2 From Boom to Slump and Back Again: The Changing
   Structure of the British Home Ownership Market   17
   Introduction   17
   The Evolution of the UK Home Ownership Market from the
      1950s to 1989   22
   The 1989–95 Slump   33
      Arreas and Repossessions as Indicators of the Severity of
         the Slump   36
      The Impact of the Slump on the Mortgage Lending Institutions   38
      The Estate Agency Débâcle   41
   The Recovery of the Housing Market   43
   What Triggered the Booms?   46
   Appendix A: Problems of Measuring Price Changes   48

3 Who Gets to Own? The Changing Social Basis of
   Home Ownership and its Implications   51
   Introduction   51
   Government Policy Support for Home Ownership   52
   The Changing Social Base of Home Ownership   55
   Home Ownership and Social Class: The Changing Relationship   57
   Who Owns What?   61
   Home Ownership, Race and Gender   64
   Home Ownership, Accumulation and Class Formation   65
   Home Owners and Political Alignment   67

4 Winners and Losers: Housing as an Investment   73
  The Calculation of Gains and Losses   76
  Distribution of National Gains and Losses: The British Household
      Panel Survey   78
  The Distribution of Gains and Losses in South East England   81
  The Determinants of Gains and Losses in the South East   84
    The Distribution of Gains and Losses by Socio-Economic Group
        and Income   86
    Gains by Estimated Current Value and Household Income   90
    Variation in Gains by Date of Purchase   92
    The Importance of Area and Current House Type   93
    The Relative Importance of Social Class, Purchase Date and
        other Variables in the Determination of Mean Gains   95
  Conclusions   99

5 Who Gets What? The Distribution of Home Owners'
  Equity in Britain   101
  Introduction   101
  Home Owners' Equity and Housing Wealth in Britain   101
    Home Ownership and the Reduction of Wealth Inequality   103
    The Distribution of Housing within Personal Wealth Holdings   106
    The BHPS Data on Housing Wealth   107
  Distribution of Housing Wealth   108
    Distribution of Current House Values among Home Owners   108
    Distribution of Housing Wealth among all Home Owners   108
    Distribution of Housing Wealth among Outright and
        Mortgaged Owners   110
    Variations in Mean Home Owners' Equity   110
    Mean Equity among Outright and Mortgaged Owners   113
  Home Owners' Equity in South East England:
      The MORI Survey   114
  The Growth and Importance of Negative Equity in the 1990s   115

6 Housing Inheritance and Equity Extraction from
  Home Ownership   123
  Housing Inheritance in Britain   123
  Measuring Housing Inheritance: The Inland Revenue Statistics   126
  A Nation of Inheritors? Who Inherits House Property   129
  Explaining the Shortfall in the Scale of Housing Inheritance   131
  Equity Extraction   133
  Measuring the Scale of Equity Extraction   137

7 Housing Careers and Housing Investment Strategies   145
   Housing: Investment or Consumption Good?   145
   Motives for House Purchase in the South East   150
   Housing Careers and Strategies   152
   Snakes and Ladders in the British Home Ownership Market   153
   Housing Ladders in Five Areas of South Eastern England   156
   The Number of Homes Owned by Occupational Class and Income 160
   Direct Measure of Movement up the Housing Ladder   161
   Conclusions   165

8 Home Ownership, Housing Wealth and
   the British Economy   167
   Housing Wealth and Consumer Spending as a Cause of
      the 1980s Boom   168
   The Home Ownership Market as a Cause of the Recession?   170
   Home Ownership and the Wider Economy: An Assessment   172
   Housing Wealth and Consumer Spending: The Contrary View   174
   The Role of Financial Deregulation in Promoting the 1980s
      Boom and Bust   178
   The Housing Market and the Economy: A Problem caused by
      the South East?   180

9 The Future of the Home Ownership Market in Britain   183
   Forces for Stability   185
      The Maturing of the Home Ownership Market   185
      Demographic Change   185
      A New Era of Low Inflation   190
      The Erosion of Confidence in the Stability of the Home
         Ownership Market   190
      The One-Off Impact of the Liberalisation of Housing Finance
         in the Early 1980s   191
   Forces for Instability   191

   References   199
   Index   217

List of Figures

Figure 1.1   Heath cartoon "Great Bores of Today . . . the market's
              gone crazy . . ."   3
Figure 1.2   Heath cartoon: "Great Bores of Today . . . we bought
              this house for . . ."   4
Figure 1.3   "Tony Blair made £200,000 here"   10

Figure 2.1   National average house prices in real terms, 1943–88   18
Figure 2.2   Annual percentage change in nominal UK house prices,
              1956–96   23
Figure 2.3   UK house price/earnings ratio, 1956–96   23
Figure 2.4   Percentage annual change in house price inflation,
              London and the North 1970–96   24
Figure 2.5   Annual percentage change in real UK house prices,
              1956–96   26
Figure 2.6   Annual percentage change in nominal and real UK
              house prices, 1956–96   26
Figure 2.7   Regional average house prices relative to the UK
              average, 1969–96   27
Figure 2.8   Mortgage advance as percentage of price paid, first time
              buyers, 1969–94   28
Figure 2.9   Annual percentage change in average UK house prices
              and average earnings, 1956–96   28
Figure 2.10   Annual percentage changes in average earnings and real
              personal disposable income, 1956–96   29
Figure 2.11   Initial mortgage repayment as percentage of average
              income, first time buyers, 1969–95   30
Figure 2.12   Loans for house purchase (net advances) by type of
              lender, 1980–96   31
Figure 2.13   Number of residential property transactions in England
              and Wales, 1978–94   31
Figure 2.14   Number of residential property transactions by region,
              1987–92   33
Figure 2.15   Mortgage arrears and repossessions, 1981–97   37
Figure 2.16   "You have been keeping up the mortgage repayments,
              haven't you?"   40
Figure 2.17   "I could have sworn I heard the all-clear"   44

List of Figures

Figure 3.1   Home ownership and party politics   70
Figure 3.2   "New Civilisation Discovered"   71

Figure 4.1   Percentage distribution of gains and losses, 1991   79
Figure 4.2   Percentage distribution of nominal illusionary gains and
             losses by region, 1991   80
Figure 4.3   Nominal illusionary gains and losses on first and current
             homes   82
Figure 4.4   Nominal crude gains and losses on first and current
             homes   82
Figure 4.5   Comparison of gain measures by year of purchase, first
             property   93
Figure 4.6   Nominal illusionary gains by year of purchase and SEG,
             current property   96
Figure 4.7   Nominal illusionary gains by year of purchase and SEG,
             first property   97

Figure 5.1   The distribution of different assets among wealth
             holders   105
Figure 5.2   Residential property as a percentage of identified
             personal wealth, 1990, 1991, 1993   106
Figure 5.3   The percentage distribution of assets in identified
             personal wealth, 1993   107
Figure 5.4   Estimated and adjusted value of houses, 1991   109
Figure 5.5   Distribution of home owners' equity, 1991   109
Figure 5.6   Distribution of housing equity of outright and mortgaged
             owners, 1991   113

Figure 6.1   Percentage distribution of assets by size of estate,
             1993–94   129
Figure 6.2   Net housing equity withdrawal, 1978–90   140
Figure 6.3   Gross equity extraction by type   141

Figure 8.1   Two simplified models of home ownership and
             consumer spending   177

Figures 1.1 and 1.2 reproduced courtesy of *Private Eye*

List of Tables

Table 2.1   The relationship between house prices and earnings   20
Table 2.2   Residential transactions by region, 1986–91   34
Table 2.3   Comparison of mix-adjusted and unadjusted prices in
             London, 1989–93   35
Table 2.4   Regional house price changes in the UK, 1983–97   35

Table 3.1   Tenure by SEG of head of household   58
Table 3.2   Socio-economic group and economic activity status of
             head of household by tenure, 1990   59
Table 3.3   Supplementary benefit recipients by tenure, 1967–87   60
Table 3.4   Estimated current mean value of current
             property by SEG of head of household   62
Table 3.5   House-type by SEG of owner occupiers, 1991   63

Table 4.1   The range of potential gain measures   77
Table 4.2   The proportion of households experiencing losses on the
             basis of their current or first property   83
Table 4.3   Mean nominal gains by SEG of head of household   86
Table 4.4   Mean nominal illusionary gains by estimated market
             value   87
Table 4.5   Type of property by occupational class   88
Table 4.6   Illusory gains, first purchase, by number of homes
             owned   88
Table 4.7   Mean numbers of years owned by SEG   88
Table 4.8   The distribution of real gains by SEG   89
Table 4.9   Nominal percentage gains by SEG, current and first
             property   90
Table 4.10   Mean absolute nominal gains by current value   90
Table 4.11   Mean nominal absolute gains by current household
             income   91
Table 4.12   Mean nominal illusionary gains by year of first purchase
             (current and first property)   93
Table 4.13   Mean illusionary gains by age of principal earner   94
Table 4.14   Nominal absolute mean gains by area   94
Table 4.15   Mean absolute gains by current property type   95
Table 4.16   Variations in mean absolute illusionary gains by SEG of
             major earner and time of first purchase   96

Table 5.1   The importance of different assets in gross and net
             personal wealth, 1994   102
Table 5.2   Percentage shares in the distribution of personal wealth
             in Great Britain   104
Table 5.3   Mean equity (estimated and adjusted) of outright and
             mortgaged owners by region   111
Table 5.4   Mean equity by SEG   112
Table 5.5   Mean equity by age of head of household   112
Table 5.6   Estimated number of home owners with negative equity   115

Table 6.1   Projections of the scale of housing inheritance:
             non-spouse transfers   126
Table 6.2   The number and value of estates passing at death   128
Table 6.3   The incidence of household housing inheritance by SEG
             and tenure, 1991   130
Table 6.4   Housing transactions with no intermediate equity
             withdrawal   136
Table 6.5   Housing transactions with intermediate equity withdrawal   137
Table 6.6   Net housing equity withdrawal   139
Table 6.7   Summary of gross housing equity withdrawal   142

Table 7.1   The importance of profit to owners in Leicester and
             Dayton, 1975   148
Table 7.2   Type of property by area   157
Table 7.3   The distribution of current market value by area   157
Table 7.4   Mean house price by type and area   158
Table 7.5   Number of rooms by property type   158
Table 7.6   Number of homes owned by study area   159
Table 7.7   Number of homes owned by age of principal earner   160
Table 7.8   Number of homes owned by occupational class of
             principal earner   161
Table 7.9   Number of homes owned by household income   161
Table 7.10   Current property type by first property type
             (2 homes only)   162
Table 7.11   Current property type by previous property type
             (3+ homes)   163
Table 7.12   Current property type by previous property type
             (3+ homes)   163
Table 7.13   Current property type by first property type (3+homes)   164

Table 9.1   Births in the UK in selected years   187
Table 9.2   Estimated number of house purchases for owner
             occupation in the UK, 1971–93   187
Table 9.3   The changing size of the 23–25 year age groups for
             selected years, UK   188

Acknowledgements

I would like to thank the Economic and Social Research Council for their financial support (grant R000233005) of the research project on the social and financial effects of home ownership in the South East of England, which was part of a linked series on research projects on the social and economic form and consequences of growth in the South East which were undertaken at the Open University. Colleagues in the South East programme at the OU – Doreen Massey, John Allen, Alan Cochrane, Linda McDowell, Phillip Sarre and Nick Henry provided a valuable forum for debate. I owe a particular debt to Phillip Sarre who also worked on the project with me and to our research assistants Beverley Mullings and Jenny Seavers. The working papers on home ownership which were produced as part of this project are listed in the references.

I would also like to thank the Rowntree Foundation for their support of the research project "Home ownership, housing wealth and wealth distribution in Britain" which formed part of a wider research programme on income and wealth distribution in Britain. Jenny Seavers played an invaluable role as the research assistant on this project.

Finally, I would like to thank Nuffield College Oxford and the Netherlands Institute for Advanced Studies for the fellowships they extended to me and the Open University and King's College London for the research leave which allowed me to work on this project. Martin Frost at King's had been an invaluable source of help and advice on a variety of economic and quantitative issues. He bears no responsibility for my mistakes. Over the years Alan Murie, Ray Forrest, Peter Saunders and Peter Williams have also all provided valuable insights and discussions on home ownership matters.

Chris Hamnett, King's College London, January 1998

Abbreviations

Conv.	Converted flat
CP	Current property
E Anglia	East Anglia
E Mid.	East Midlands
FP	First property
Man.	Managerial
NM	non-manual
NW	North West
OMA	Outer Metropolitan Area
ONM	Other non-manual
OSE	Outer South East
P/B	Purpose-built flat
ppc	percentage point change
Prof.	professional
ROSE	Rest of the South East
SE	South East
SEG	socio-economic group
Semi	semi-skilled
Semi-det	semi-detached
SW	South West
Unsk.	unskilled
W Mid.	West Midlands
Yorks/H	Yorkshire and Humberside
YoY	year on year

(1 billion = 1,000 million throughout).

CHAPTER 1

Home Ownership in Britain: Riding the Roller Coaster

The growth of home ownership since the end of the Second World War marks one of the most fundamental social changes to have taken place in Britain. From being a nation of renters at the end of war, Britain has been converted into a nation of home owners. In 1945 approximately 25% of households in Britain owned their own homes. Today the proportion is just over two-thirds. In the process, the proportion of households renting from private landlords has fallen from 65% to about 8%. As a result, the home ownership market in Britain plays a far more important role today than hitherto: both in housing the population and as a potential source of capital gains and losses. In addition, the home ownership market plays a significant role in the overall health of the economy. This is not to deny the importance of social and private rented housing or the major problems of homelessness. It is simply to assert that the home ownership market now affects two out of three households in Britain, and many more who wish to gain access to it.

This book is about the dramatic booms and busts of the home ownership market in Britain during the last twenty years: and their causes and consequences both for the individuals involved and for the economy as a whole. It argues that the home ownership market in Britain, particularly in southern Britain, where the booms and slumps have been experienced most sharply, has been akin to a casino. There have been big winners, but there have also been big losers. The last thirty years have been a roller coaster ride for owners: exhilarating, but potentially highly dangerous, not least for those who fell off, or were thrown off, in the slump of the early 1990s.

The Booms of the 1970s and 1980s

The origins of the modern British housing market can be traced back to the 1960s, when Britain experienced a decade of slow but sustained house price inflation which doubled house prices. This was followed in the early 1970s by the first of three major postwar booms which doubled house prices once again. Then came another boom in the late 1970s, and the long boom in the mid-late 1980s. Not surprisingly, home ownership came to be seen as an insurance against inflation and as a source of capital gains. This experience coloured the attitudes and expectations of a generation of home owners in Britain. From the mid 1960s

1

onwards, if not before, buying a house was considered a sound investment, possibly the best anyone could make. As inflation ticked away, the real value of the debt was progressively reduced, while the value of the asset increased. By the time the mortgage was paid off, owners were sitting on a substantial capital asset. Home ownership functioned as the principal vehicle of domestic capital accumulation for millions of people. From this perspective houses were not simply places to live in or roofs over our heads: they also functioned as important investment vehicles.

As Doling and Ford (1991) noted: "A feature of the housing market in Britain is the enduring belief that home ownership is one of the best, if not the best, investment accessible to ordinary people". House prices and house price inflation were a common topic of conversation in London and the South East (providing a source of radical embarrassment or smug complacency depending on political outlook). Cartoons by Alex, Heath and others (Figures 1.1 and 1.2) highlighted the mood of the time. For a small minority, playing the housing market through a sequence of upward moves, renovations and resales became, rather like the stock market, a source of both satisfaction and profit. For most people, however, the housing market was something that provided financial benefits in addition to the benefits of residence. There was a fantasy or fairy tale element to it all, rather like medieval alchemists turning base metals into gold. At the height of a boom the value of your house could easily rise by more than the average annual salary.

Not surprisingly, some people treated these gains like fairies' gold, borrowed against it and spent it. Then, just as in the Grimm fairy tales, the debts were called in. As Pawley (1978) tellingly described it, the home ownership market became

> a game of reverse monopoly whereby, instead of using money to buy houses to put on your street, you took money out of the houses.... As house prices rose, so the housing market became a massive source of equity withdrawal for many owners, with huge sums being extracted each year in the form of loans, cash release mortgages, trading down and the like, all funded by the mortgage lenders who were only too keen to lend on the security of a house.

At the peak of the 1980s boom prices in London and the South East spiralled up to what seemed to be (and were) astronomically high levels, and media attention was directed to a handful of sales which were emblematic of the house price mania which gripped London. Two stories in particular captured media attention. The first, in February 1987, concerned the offer of £36,500 made for a converted 11' by 6' broom cupboard in a prestige block of apartments, Princes Court, opposite Harrods. The "flat", which also included a 2' 6" square lavatory, washbasin and shower area, as well as fold-up bed (but no cooking facilities), was advertised as an ideal *pied-à-terre*. This sale roused the ire of Tom Torney, Labour MP for Bradford, who said: "Anyone who is foolish enough to pay that price for the sake of a swanky address is merely inflating still further the ludicrously high

2

Great Bores of Today

"... the market's gone crazy I mean did you read about that cupboard in Knightsbridge which went for $95,000 and yet last week we were up in Cumberland seeing Penny's mother and we saw this amazing old priory which has been converted into a hotel it's got 18 rooms, 29 acres, stables and outbuildings and a small cottage in the grounds and it's going for 12 grand you could proba ly get it for less if you give them cash have you seen the latest copy of *Country Life* it's incredible what you can pick up Grade One listed buildings, Georgian stately homes, landscape gardens all for half the price of a garage up here I mean I can easily fax my stuff in and there's an airport in Huddersfield which apparently does a shuttle I mean we're seriously thinking about moving out and you've got to get in fast Gloucestershire's peaked already it's the same story in Dorset if you want to make big money I'd start buying up the border country before it goes the same way we saw a castle going for 3 grand ..."

Figure 1.1   Heath cartoon "Great Bores of Today ... the market's gone crazy ..."

Source: Private Eye, 663, 1 May, 1987

3

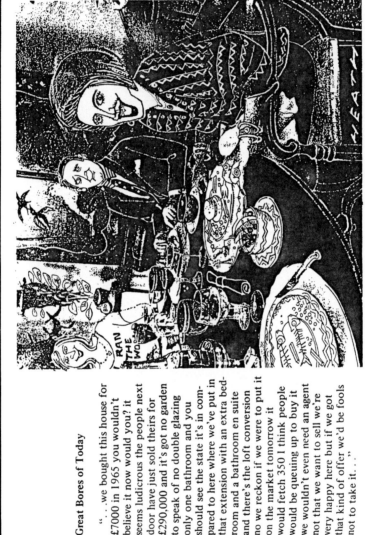

Great Bores of Today

"...we bought this house for £7000 in 1965 you wouldn't believe it now would you? it seems ludicrous the people next door have just sold theirs for £290,000 and it's got no garden to speak of no double glazing only one bathroom and you should see the state it's in compared to here where we've put in that extension with an extra bedroom and a bathroom en suite and there's the loft conversion no we reckon if we were to put it on the market tomorrow it would fetch 350 I think people would be queuing up to buy it we wouldn't even need an agent not that we want to sell we're very happy here but if we got that kind of offer we'd be fools not to take it..."

Figure 1.2 Heath cartoon: "Great Bores of Today...we bought this house for..."
Source: Private Eye, 643, 8 August, 1986

4

prices of property in London. For that money you could get a four-bedroomed house, with all mod cons and a jolly good garden in Bradford" (*Independent*, 1987).

The second case was in June 1987, when a one-room freehold studio house complete with a tiny kitchen and lavatory but no bathroom, in Wilby Mews, Holland Park, was sold for £58,000 (*Times*, 2 June 1987). Judging from the pictures it appeared to be a converted garage or single-storey extension. A variety of stories focused on the growing north–south divide in house prices and pointed out that for the price of a very small flat in central London it was possible to buy a country house in northern Britain. Such comparisons are a staple of the early years of a boom when prices have risen sharply in London and the South East but remain depressed in most of peripheral Britain. Such large disparities are a common feature of the British home ownership market in boom times and, given the geographical disparities in prices between the affluent South East and other parts of Britain, they are almost inevitable. It is usually forgotten that they narrow again during downturns when prices continue to rise slowly outside the southern regions.

The 1990s Slump

In the event, the expectations of many owners and buyers in the mid to late 1980s were dashed by the slump in the early 1990s. This started in London and the South East in autumn 1989, when the house price boom burst like a pricked balloon as a result of the rapid rise in interest rates introduced by the then Chancellor of the Exchequer, Nigel Lawson, in a desperate attempt to reduce escalating consumer spending and inflation. There had been slumps before in the British home ownership market, notably from 1974 to 1977 in the aftermath of the early 1970s boom and from 1980 to 1981 in the wake of the late 1970s boom. But, although turnover dropped sharply in both these slumps, prices remained static in money terms even though they fell sharply in real terms. Consequently, relatively few owners were forced to sell at a loss in money terms (even though they may have had a loss in real terms). In the early 1990s slump, however, prices fell sharply in cash terms and many new buyers and some existing owners found themselves with substantial losses. Between 1.5 and 2 million home owners were thought to have negative equity, that is, the value of their house was less than the outstanding mortgage, and over 400,000 homes have been repossessed by lenders during the 1990s when their owners failed to meet their repayments. In London and the South East (where the 1980s boom had been most marked) prices fell by 30% in cash terms and by far more in real terms. This proved a dramatic shock to home owners and lenders alike, accustomed as they were to steady house price inflation.

Prior to 1989, owners' experience of the home ownership market in Britain had been largely positive. While some leasehold flat owners in London had suffered at the hands of rapacious freeholders or agents (Hamnett and Randolph, 1987) and some council tenants who had bought their flats under the "Right to Buy" provisions had found that they could not sell so easily, for most people home ownership was an insurance against inflation and a source of capital gains. The prolonged housing market slump of the early 1990s radically changed this set of expectations. In London and the South East (where the 1980s boom was greatest), where the fall in prices and the rise in negative equity was much sharper than elsewhere, large numbers of young first-time buyers drawn into the buoyant housing market in the late 1980s had their fingers badly burnt. For this group, the shift in outlook and confidence was dramatic. The home ownership market was not a source of capital gains, but a source of financial losses, and in some cases of debt and repossession. As Forrest and Murie (1994: 55) put it:

> The inflationary discourse of trading up, gentrification, gazumping, and equity gain has given way to a new language of negative equity, debt overhang, repossessions and arrears. In the late 1980s an overheated, over-indebted, hypermobile housing market rapidly degenerated towards a gridlock of immobile households adjusting to very different expectations of home ownership.

The losers were those who had bought their house fairly recently at, or near, the peak of the market and who faced a paper loss of some tens of thousands of pounds or even more in some cases. This did not matter too much if you kept your job and income and were able to keep paying the morgage. In such cases, the loss was simply a paper one. The problems occurred where people had lost their jobs, split up or divorced and had to sell their house. Then the loss was converted from paper to a real one with damning consequences. Perhaps the most emblematic case of the 1990s slump concerned a London based interior designer, who borrowed £630,000 from the Town and Country building society to purchase a three-storey terraced house in Hampstead. When his business ran into difficulties after just two payments, he applied to the DHSS for assistance in paying his mortgage and they agreed to meet the interest costs of £7,000 a month. When it emerged that he had defaulted on a previous loan, the payments were stopped and the Town and Country was forced to repossess. The house sold for less than £450,000 and, with accumulated arrears, the Town and Country was left with a £300,000 bill. This was by no means the society's only problem loan and it was subsequently taken over by the Woolwich with provisional debts of £42 million in a successful bail out operation (Rathbone, 1992).

As the slump dragged on and sales stagnated and prices continued to fall, the focus of media attention shifted: commentators began to ask whether the slump would ever end and whether the home ownership market was sliding into an abyss. As Rachel Kelly (1992: 34) wrote in *The Times*: "No one believes that house prices will boom any more. The bonanza of the 1980s was a one-off. When prices finally start to rise, they will do steadily and in line with inflation.

Gone are the days when you could make more money by listening to the experts and waiting for their predictions of house price rises to come true." And John Eatwell (1992) writing in the *Observer* commented that:

> it looks increasingly likely that the decline in house prices will become self-perpetuating. As prices fall, it is rational for buyers to postpone pur-chases, while sellers desperately try to sell as quickly as possible. With little new demand coming into the market, the overall price level of the housing stock is held up only by what people believe houses are worth. It only needs every-one to believe that houses are worth 10 per cent less for them actually to be worth 10 per cent less.

Eatwell was right. Houses are worth what someone will pay for them and, in a slump, what people are prepared to pay can fall sharply as expectations fall. But memories are often too short. When a boom is under way it seems incon-ceivable that it could end, and during the depths of the slump some thought that the market would never recover, that an era had ended and that housing market booms were a thing of the past. Some have longer memories. In a perceptive article, "Nightmare on Acacia Avenue", written in the depths of the slump, Gail Counsell (1992) noted the sense of fear which permeated the market and identified what she termed the "doomsday scenario" where the house price slide turns into free fall: "As house prices have tumbled across a swathe of southern England, dismay has increasingly turned to panic. Home owners and the housing industry have begun to feel as though they are gazing into a bottomless abyss, and the Government is under pressure to do something to shore up the market." But, as Counsell pointed out, "There are precedents for what we are seeing, and on the evidence of those precedents there is every reason to assume that the market has a bottom. We are simply not there yet." Pointing to previous booms and busts, she argued that house prices might not recover until 1994 and that prices, which had fallen by around 6%, would have to slide a further 5–10% to restore equilib-rium. She firmly rejected the doomsday scenario: "For prices to fall below three times earnings, there would have to have been either national economic collapse or a radical shift in the desirability of owning property. In the absence of an economic slump comparable with the 1930s – and we are a long way from that – it is difficult to imagine what could provoke such a shift."

In the event, prices bottomed out in 1993 and continued to bounce along the bottom for another three years, until they slowly began to rise again between the end of 1995 and early 1996. Rather than rushing to sell at the depressed prices, potential sellers, believing that their house was worth £80,000 rather than the £60,000 the market said it was worth, or perhaps unable to take the loss, simply took their houses off the market and sat tight. The home ownership market in Britain is like other asset markets in that it is essentially cyclical with a consist-ent pattern of boom and busts. Although the 1990s bust was the most spectacular to date, and the most severe in its impact, it did not mean the end of house price inflation. Rob Thomas argued in 1996 for instance that, with house prices low

7

relative to incomes and low mortgage rates, the conditions were in place for another cyclical upswing.

The Recovery

Events have already proved Rachel Kelly wrong and Gail Counsell correct. At the time of writing this chapter, the British home ownership market is climbing out of its deepest recession in fifty years, if not since the 1920s. House prices have ceased falling as they did through the early 1990s, and have begun a steady recovery, rising by 10% in 1997 and almost double this in London and the South East. Indeed, the London-based media have begun to use the term "boom" once more, particularly to describe the housing market in some parts of inner London with easy access to the City, where prices have risen sharply.

There is nothing the British press (and seemingly the public) loves more than a good housing boom, with all the thrills of big capital gains and the anguish of hopes dashed for potential first-time buyers who see prices spiralling out of reach. All the stuff of human life is there, or so it seems. There are winners and losers in the Great British housing lottery and all you need to do to take part is to buy a house and hope for a boom. The press coverage after the doom and gloom of the early 1990s is illuminating: "Housing market in best shape since Eighties" *Independent*, 1 June 1996; "The housing market takes off again" *The Times*, 21 June 1996; "Boom, boom, crash – Is this the echo of 87?" *Observer*, 23 June 1996; "House prices to rocket 10% in 'mini-boom'" *Independent*, 1 July 1996; "Lift-off for house prices" *Evening Standard*, 10 July 1996; "Up but not through the roof", *Observer*, 25 August 1996; "Property poised for blast-off" *Independent on Sunday*, 15 December 1996; "Will house prices hit the roof again?" *Independent*, 28 December 1996; "Another housing boom?" *The Economist*, 18 January 1997; "Housing recovery should pick up speed" *Independent*, 21 February 1997; "House prices soar in London and the shires" *The Times*, 12 June 1997; "The biggest boom since the Eighties" *Independent*, 28 August 1997; "Boom is back with a vengeance: As house prices soar in the capital there's money to be made" *Daily Telegraph*, 20 September 1997.

The tone behind the headlines varies from the cautious to the near exultant. "Boom, take-off, blast-off, soar, rocket": the words almost make you feel that we are witnessing a Grand Prix or a blast-off from Cape Kennedy, and almost without exception, the message from the headlines is that a boom in the housing market is what everyone has been waiting for. The text of some of the articles qualifies this enthusiasm, and points to the dangers of another boom, but the enthusiasm is understandable in the context of the slump of the early 1990s and the severe doubts which were expressed as to whether the home ownership market would ever recover. In some parts of London prices have taken off again in a way reminiscent of the 1980s. "Gazumping" has re-entered popular vocabulary as buyers have begun to bid up prices to get the property they want or sellers have started playing off one buyer against another to maximize sale price.

Stories have begun to circulate in the newspapers of central London properties going for astronomical sums. As Anne Spackman (1997) ironically commented in the *Financial Times*:

> Two "compact apartments" have come up for sale in Knightsbridge, London. Described as south-facing with a view and access to communal gardens, they are actually the converted landings of a house in Thurloe Square, measuring 10ft 5in by 8ft 6in each. Priced at £62,000 apiece they carry a £700 annual service charge. The last time converted Knightsbridge broom cupboards were hot properties was in 1987, a few months before the residential property bubble began to sag. Does this year's crop herald similar trouble in 1998?

Spackman argued that it did not, on the grounds that the boom was still confined to London and the South East, where prices have risen rapidly. But she has a long memory, and an awareness that the home ownership market in Britain during the 1970s and 1980s was strongly cyclical. So does the economist Roger Bootle (1997), who suggested in *The Times* that there are two reliable indicators of the top of the housing market. The first is when middle-class couples talk of having to buy a property for their children for fear that they will never be able to afford one themselves. The second: "is the appearance of the comet-like story that a former broom cupboard in Knightbridge has just fetched a good price as a mini studio flat". Bootle ironically suggested that, as broom-cupboard stories have resurfaced, that settled it for him: "next year, the rate of house price inflation would be lower and in central London we might see absolute price falls".

Certainly, prices in London and the South East recovered strongly in 1996–7. An article in *The Times* (7 March 1997) titled "Home sellers earn bonus from 'golden postcodes' stated that golden postcode areas are emerging in Britain where the number of buyers far exceeds sellers and it is almost impossible to find a house." It reported that: "Earlier this week a three-bedroomed cottage in Battersea south London went on the market for £185,000 and in 6 hours 21 people had viewed it and four offers of the asking price were submitted. It sold the following day for more than £200,000 after sealed bids were made." The article went on to quote estate agents as saying that "the bonus culture in the City has pushed up Islington house prices. . . . Islington is a historically under-valued area and the bankers are pouring money into it." House prices are soaring in Barnsbury, Tony Blair's Islington neighbourhood. "A house in his area worth £450,000 last year would now cost £525,000 at the very least, but they are very difficult to get hold of. Once people go somewhere like that, they do not want to move anywhere else" (*The Times*, 7 March 1997).

Rather ironically, but emblematic of the recovery in the inner London market, Tony Blair's move to 10 Downing Street meant that the Blairs were forced to sell their Georgian home in Islington. *The Times* reported that the five-bedroom house, which the Blairs had bought for about £375,000 in 1992 at the depth of the housing market slump, and had spent thousands of pounds improving, had

9

Figure 1.3 "Tony Blair made £200,000 here"
Source: *Financial Times*, 12 June 1997

been put on the market for £615,000. The house sold very quickly at its asking price, giving the Blairs a paper gain of £200,000: a return of 50% over five years, and higher if the mortgage is excluded and the return is calculated just on the money the Blairs put into the house themselves. An ironic cartoon in the *Financial Times* showed the house with a commemorative blue plaque saying "Tony Blair made £200,000 here" (Figure 1.3) and the *Independent* (12 June 1997) carried a short editorial, "Blair scoops the lottery", which asked:

> Here is a question for people who believe in fairness and social justice. Is it right that someone who buys a house can make a quarter of a million pounds profit in a few years, while someone else who lives a block or two away is unable to accumulate any capital at all? This question is not a dig at the Blairs; they are simply making the kind of casually huge gain that many others do. But the sale highlights a curiosity of British society: big differences in wealth (and thus comfort and security) often have more to do with fortuitous house purchase, where you moved and when, than any other factor. Yet politicians rarely contemplate making our housing market more rational, because then the lottery might end.

I quote the *Independent* editorial in full because it hits the nail on the head. Big differences in housing wealth in Britain do often have more to do with what you buy, and when, than with any other factor. The home ownership market in

Britain has functioned as a massive, though far from random, lottery, distributing differential gains and losses to millions of owners across the country and over time. It is far from random because there is a broadly consistent pattern of gains and losses depending on type of property bought, where and when it was bought, and who bought it. Again, the Blairs' case is emblematic. If we wanted to construct an Identikit picture of the typical big winners in the British housing lottery we need look no further. The house is large, expensive and located in inner London and the Blairs are a young professional couple with a large income by most people's standards. Cherie is reputed to earn around £200,000 a year, and Tony Blair earned around £30,000 a year as an MP when they bought the house. The house was sold to a French couple in their thirties: the husband works in the City and wanted to be close to work. Who else but a banker or dealer in the City, a successful lawyer, doctor, entrepreneur or overseas buyer, could afford to pay £600,000 for a house when, according to the Nationwide Building Society, the average price of a detached house in suburban London in the second quarter of 1997 was £172,000 and the average price of a terraced or semi-detached house was fractionally over £100,000?

The case of the Blairs' house is not isolated or idiosyncratic. It is a story which was repeated thousands of times in expensive areas of London and millions of times across Britain during the 1970s and 1980s, though the sums involved were generally much smaller. The point is that, had the Blairs bought a cheaper house, had they bought outside fashionable inner London, or at the end of the 1980s rather than the early 1990s, their gains would have been much lower. Indeed, we could have read the story of the Labour Prime Minister who lost money selling his house. The headlines in the tabloids can be imagined: "If he can't make money on his house can he run the country?" or perhaps more likely "New Labour, Sad Loser". Given the ubiquitous nature of southern price falls in the early 1990s it is very possible that the Blairs lost money on their previous home if they bought in London in the mid to late 1980s. This would merely illustrate my point that timing and location are important in the capital gains and losses lottery.

My wife and I moved to London in 1970 to rented flats in Belsize Park and eventually managed to buy a flat in West Hampstead in late 1993, having already been gazumped on several occasions. Mortgage rates and inflation soared, and real (inflation-adjusted) house prices had fallen by 30–40% by the time we sold in late 1976. Although we sold for £1,500 more than we had paid, we took a massive loss in real terms. Fortunately, we bought a large terraced house in north London whose price was equally depressed at the time, but whose real value has since risen considerably. We were lucky. If we had been a generation later, and bought our flat in 1987 and sold in the early 1990s, we would have found ourselves with large negative equity, quite unable to move, let alone buy a house. In the event, the cash value of our house fell 30% from a peak in 1988 to the early 1990s, subsequently almost doubling in value by 1997. The fall in real terms during the early 1990s, of perhaps 50%, was not a problem, first because we did not move, and second because by 1993 our mortgage was much less than the current value of the house as a result of inflation since 1976. Others have not been so fortunate. They entered the housing market at the wrong time at or near the peak of a boom (as we did in the early 1970s) and then experienced large nominal price falls which left them financially exposed.

Key Arguments of the Book

This book does not attempt to provide a history of housing policy in Britain. This has been comprehensively done by Holmans (1990), Malpass and Murie (1990) and others. Nor does it attempt to provide a general overview of owner occupation, which has been done by Merrett and Gray (1982), Forrest et al. (1990), Pawley (1978) and Daunton (1987) amongst others. My focus is more specific and concerns the nature of the home ownership market, the determination of house prices and the distribution of housing wealth, capital gains and losses. It is about the scale, distribution and characteristics of winners and losers in the home ownership market in modern Britain, and the relations between the housing market and the wider economy. The book has a number of key arguments:

- The home ownership market is highly cyclical, with booms and slumps and a distinct geographical pattern. Booms invariably start first and finish first in London, and spread more or less rapidly to the rest of the country.
- House prices are closely related to changes in real incomes in the long term but booms are triggered by changes in the number of people in the key first-time buyer age groups, by increases in real income and mortgage availability.
- There are inevitably winners and losers in this process, and timing and area of purchase are crucial in determing the extent of the gains or losses from the housing market. The absolute size of gains is also strongly related to social class.
- Equity accumulation is highly differentiated by type and size of property, area, and date of purchase. Generally speaking, the longer a property has been owned the greater the capital accumulation. By definition, the chief losers in this process are tenants, who do not accumulate housing wealth, or those with negative equity.
- Home ownership has now become so widespread and so diverse that booms and slumps can and do affect a wide range of people, not just a few middle-class property owners. Home ownership is no longer a middle-class tenure but it is differentiated by property type and size.
- Wealth accumulation through home ownership has now become the principal form of wealth accumulation for the great majority of the population notwithstanding the impact of the slump and negative equity. Housing wealth accounts for over 40% of net personal wealth in Britain.

12

- A substantial amount of housing wealth is transmitted via inheritance to the next generation. Housing is the key element of inheritance but it has not grown as rapidly as predicted ten years ago.
- Equity extraction from the housing market is considerable and can exert a substantial influence on consumer spending when prices and housing wealth are increasing rapidly.
- The links between the housing market and the economy have now become so strong that a sharp upturn or downturn in the housing market can have strong adverse effects on the economy as a whole, although it would be wrong to blame the housing market for government economic mismanagement and inflationary booms.
- Although some commentators have suggested that the 1990s slump and the advent of an era of low inflation mark the end of the house price booms of the 1970s and 1980s, this is unlikely. We are likely to see another boom within five years.

A number of important themes and questions inform the debate in this book. These concern the cyclical process of boom and bust, the changing social character of home ownership and the basis it provides for class formation, the distribution of winners and losers from home ownership, the impact of home ownership on wealth distribution and inheritance, and the impact of boom and bust in the home ownership market for the economy as a whole. Each of these debates will be briefly considered in turn.

As stressed earlier, the late 1980s boom and the early 1990s bust were not the first, and are not likely to be the last, booms and busts in the British home ownership market. Such booms and busts are common to housing markets in most capitalist countries; they have a strong cyclical pattern and a distinct geography. The history and character of the boom and bust is considered in detail in Chapter 2 along with an analysis of their causes. Until 1970 the majority of households in Britain were tenants and ownership was largely restricted to the middle classes. The postwar period has seen the diffusion of home ownership across the social scale with major implications for wealth accumulation, political affiliation and management of the economy.

Economists' interest in home ownership was fuelled in the 1960s and 1970s, when it was realised that the postwar growth of home ownership and rising house prices might be playing a significant role in the redistribution of wealth in Britain. Previously relatively few people had owned assets of any significance, and property ownership was concentrated in the hands of a relatively small minority of landlords: Galsworthy's "men of property". But home ownership widened property ownership and popular wealth, with important implications for social stratification and political divisions, as Chapter 3 shows.

Much of the sociological debate concerning home ownership as a source of capital gains and wealth accumulation was triggered by an important paper by Peter Saunders (1978). Saunders argued that house prices in Britain had risen faster than the general rate of inflation since at least the end of the war, had risen over the preceding ten years faster than returns on any other form of investment, and showed no signs of abating. He argued that home ownership thus provides "access to a highly accumulative form of property ownership which generates specific economic interests which differ both from those of the owners of capital and from those of non-owners". He went on to assert that these interests provided the basis for a specific class formation distinct from that of the owners of capital and non-property owners. This argument proved a red rag to many academic commentators, who fiercely disputed the empirical validity of the argument and its implications for class formation. In a nutshell, the idea that home ownership could provide the basis for distinct social classes, separate from the classes formed in the world of production and labour markets, was anathema to many sociologists. Saunders (1984) added fuel to the flames by claiming that what he termed "consumption cleavages" could be more important than the traditional bases of social class. If Saunders's claims were correct, then much of the sociology of social class could lie in ruins. Not surprisingly a veritable industry developed examining the evidence for and against Saunders's thesis, which is discussed in Chapter 3.

Saunders's work is extremely important for social theory but, as the housing slump of the last few years has clearly highlighted, it is empirically suspect, at least in the short term. In the long term, he may be correct about the accumulative potential of home ownership, not least because housing has a high and positive income elasticity of demand. In other words, for a given increase in income, people are likely to devote a high proportion to housing. House prices have tended to move in line with incomes rather than retail prices, and consequently the capital gains from home ownership represent what Saunders terms "a stake in the system". The gains are not guaranteed, however, and it is clear that there are both winners and losers from the home ownership market. Not everybody is a winner and some people can and do make considerable losses. Therefore his theoretical claims about the accumulative potential of home ownership as a general basis for a class of domestic property owners are dubious. While it may provide a basis for a variety of groups of home owners, there is no single class.

The booms and busts of the housing market, especially the dramatic impact of the housing market slump, were also thought to play a role in changing the political affiliations of voters. Margaret Thatcher came to power at least in part because of her promise to allow council tenants to become home owners by buying their homes at a substantial discount. The Conservatives hoped these people would become Conservative voters in the process, putting the final nail in the coffin of the left, who had strong support among council tenants and who viewed home ownership with suspicion as fostering petit bourgeois tendencies. But with the slump, and the impact of negative equity and repossessions on the marginal working-class home owners who had supported Thatcher, there is some evidence that disgruntled home owners may have played a part in the defeat of the Conservatives at the 1997 General Election. Saunders argued that the accumulative potential of home ownership could create the basis for a political alignment, but it appears that winners and losers may have voted very differently. Chapter 4 provides a detailed analysis of the distribution of winners and losers in the home ownership market both in Britain as a whole and in the South East using evidence from two large surveys, and Chapter 5 examines the distribution of housing wealth and capital gains.

In the late 1980s it was realised that housing wealth had major implications for inheritance. Whereas relatively few people owned wealth and had wealth to leave, the spread of home ownership meant that inheritance was likely to become a much more widespread phenomenon than hitherto. Commentators claimed that Britain was rapidly becoming a nation of inheritors. A golden age of inheritance was in the making where, to quote John Major, "wealth would cascade down the generations", aided by Conservative policy, which was designed to reduce inheritance taxation and raise tax thresholds. Chapter 6 shows that home ownership has greatly widened the distribution of wealth ownership in Britain, but the notion of a "nation of inheritors" and of housing wealth "cascading down the generations" has a number of problems, not least those of equity extraction, which grew rapidly during the late 1980s, and the increasing cost of funding residential care in old age. Chapter 7 examines the extent to which home owners have actively used the housing market as a vehicle for speculation as some observers have claimed, and the extent to which home ownership has changed from providing a roof over people's heads to being primarily a form of financial investment. I will argue that, for most people, housing still primarily functions for use and as a home rather than as an investment, although owners are now well aware of the potential gains (and losses) offered by home ownership.

Economists' interest in home ownership was also stimulated during the late 1980s when it began to be realised that the growth of personal sector wealth and the combination of equity extraction and the so-called "feel good" factor were playing an important role in driving the growth of consumer spending and inflation. The home ownership sector was now so extensive and capital gains and equity extraction so large that the housing market became a potent force in the economy and a problem for economic management. It was feared that Britain, or at least southern Britain, had become a nation of speculators with the goal of making their fortune from home ownership. Indeed, it was argued that the housing market boom in southern England during the mid to late 1980s had a detrimental effect on the national economy by pushing up inflation, leading to anti-inflationary policies which affected Britain as a whole. So, too, it was argued that house price inflation in southern England was leading to the country being split into two: an affluent, high priced South and a depressed North, with consequent effects on labour migration and regional policy. The regional house price gap was thought to be gradually widening.

It was also realised that the financial deregulation of the early 1980s played a central role in the housing boom: the banks were able to compete with the building societies for mortgage lending and both entered into a competitive struggle to gain market share by increasing mortgage advance to price ratios and relaxing mortgage advance to income ratios. As a consequence, many borrowers were able to borrow more, and more easily, sucking previously marginal buyers into the market. In addition, changes in government policy enabled mortgage lenders to lend for consumer spending and equity release if the loans were secured against the value of the property. The Bank of England sounded warnings in the late 1980s but these went largely unheeded, and a new breed of mortgage lender was attracted into the market by the promise of quick and easy profits. This dream turned sour in the 1990s as the housing slump led to mortgage lenders and insurers losing billions of pounds as hundreds of thousands of homes were repossessed. It was then argued that the slump in the housing market and the rise in personal sector indebtedness was now having a great negative impact on consumer spending and economic growth. It was argued that unless the housing market could be jump-started the economy would remain mired in recession. In both boom and recession, the housing market was now seen to be a key element in economic management. It was too important for the Government to ignore, particularly when large numbers of homes were being repossessed. In the depths of the slump, there were suggestions that the housing market in Britain was on the edge of an abyss, and there was pressure on Government and mortgage lenders to act. These issues are considered in Chapter 8.

Finally, Chapter 9 looks at the future of the home ownership market in Britain, focusing on the factors for stability and instability and on the demographic changes which have been extremely influential in underpinning the booms of the 1970s and 1980s. It asks whether there will be other booms or whether the "end of inflation" in the 1990s marks the end of the 1970s and 1980s boom.

CHAPTER 2

From Boom to Slump and Back Again: The Changing Structure of the British Home Ownership Market

Introduction

Britain has experienced three major house price booms over the past thirty years: 1971–73, 1977–80 and 1986–89. After each boom there has been a substantial market downturn during which prices fell in real terms. The slump of the early 1990s was unique, however, both for the length of the downturn (1989–95) and its severity: prices fell by 30% in London and the South East in cash terms and by far more in real terms, sales slumped, more than 400,000 homes were repossessed and over a million home owners suffered negative equity. In addition, it has administered a severe shock to confidence in the housing market, which may take years to recover fully. The longstanding belief that houses are a safe investment and a source of almost guaranteed capital gains has been severely undermined, particularly for the large number of first-time buyers who entered the market in the 1980s. The purpose of this chapter is to examine the history and structure of the cycle of booms and slumps in order to provide a context for subsequent chapters and show the regional differences in house prices and the changes in affordability and prices over time. First, however, it is necessary to address the question of why house prices rise at all, and why they rise unevenly rather than steadily.

Broadly speaking, house prices rise over time because housing is a key element of household consumption and, as incomes rise, people are both willing and able to put a substantial proportion of their income into buying a house or moving upmarket to a larger, more desirable and usually more expensive home. As already noted the number of home owners has risen from 4 million immediately after the Second World War to 16 million today. Supply is largely fixed in the short term as the building industry only builds 150,000–200,000 units a year (about 1% of the owner occupied stock) and planning restrictions mean that the supply of land is limited, particularly in southeast England, which tends to push up house prices (Evans, 1989, 1991a, 1996). Although over 6 million houses have been transferred to owner occupation from private and social renting, since the war demand for home ownership has exceeded supply. House prices thus tend to rise faster than prices in general (Marfleet and Pannell, 1996).

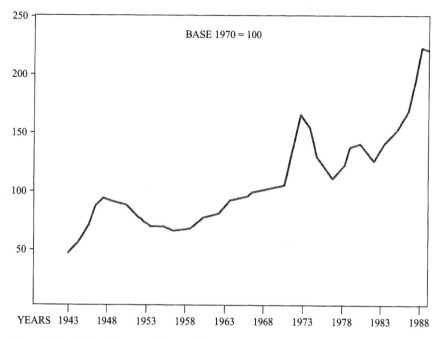

Figure 2.1 National average house prices in real terms, 1943–88
Source: Holmans (1990)

National average house prices doubled from £2,500 in 1960 to £5,000 in 1969; they rose to £65,000 in 1990, an increase of 2,500% in nominal terms. After taking retail price inflation into account, Holmans (1990) found that house prices rose 3.2 times in real terms between 1943 and 1988, and 2.3 times between 1970 and 1988 (Figure 2.1). Cutler (1995) notes that real house prices rose fourfold between 1943 and 1994, an average annual rate of 2.7%, although this does not take into account improvements in quality or changes in type and the mix of houses sold. When this is done, Duncan (1990) suggests that real price rises have been considerably less than is generally believed, though it is difficult to accept that there has not been a substantial increase in real house prices even allowing for these shifts. Cutler (1995) notes a similar upward trend in real house prices in the G7 countries, averaging 1.6% per annum, with the UK and Japan experiencing the fastest average rates of growth. She points out:

> The upward trend in house prices relative to consumer prices is explained by a combination of steadily rising demand for owner-occupied housing – a result of rising per capita real incomes and increasing population, combined with the greater availability of mortgage finance and tax advantages favouring home ownership – and a relatively inelastic supply of new dwellings because of the limited supply of land (Cutler, 1995: 261).

18

Paradoxically, however, this rise in real house prices does not mean that houses have become more expensive in relation to incomes over time, as incomes have increased much more rapidly than retail prices. The ratio of average house prices to average earnings has averaged 3.5 in Britain over the past forty years (that is to say, average house prices are 3.5 times average annual male earnings) and, as earnings rise so do house prices (Table 2.1), though not evenly (Marfleet and Pannell, 1996). At the end of the day, house prices in a market economy are controlled by ability to pay, and thus by incomes and the availability of mortgage finance.

Although the house price earnings ratio measured on a national basis has fluctuated sharply during the past forty years, ranging from a low of just under 3.0 in 1960–61 to almost 5.0 in 1973 and 4.3 in 1989, it is now back to its long-term average of 3.5. Peter Saunders (1990) suggests that this long-term stability is very significant because it means that, while house prices have risen in real terms relative to other commodities, they stayed constant in cost relative to real disposable earnings. While some entrants to the market have to pay more in mortgage payments relative to earnings than previous buyers, depending on the stage in the cycle at which they entered and the prevailing mortgage rate (particularly for those who bought between 1989 and 1992), in the long term house prices rise in line with the rise in incomes. As Saunders (1990: 145) puts it:

> Existing owners gain, not at the expense of the next generation of buyers, but by extracting a portion of the growing real wealth of the society in which they live. The result is that home ownership has come to represent the equivalent of a certificate of entitlement to share in the fruits of economic growth. It is, despite all the critics' claims to the contrary, quite literally a "stake in the capitalist system".

Not surprisingly, this claim attracted considerable criticism from the left, not least for its ideological overtones, but Saunders is essentially correct in his claim. Because house prices are related in the long term to increases in real income, there is a direct link between prosperity and house prices. This does not mean, however, as Chapters 4 and 5 show, that we have to accept Saunders's other claims that house prices always continue to rise and that all home owners gain. The experience of the early 1990s has demonstrated the falsity of that claim. In general, however, and leaving aside demographic factors, long-term trends support Saunders's contention (1990: 153) that "as long as the capitalist economy keeps growing, house prices are likely to keep rising"; the experience of the 1990s slump, however, shows that this is only true in the medium to long term. In the short term the reverse can prove to be the case, as new buyers in the late 1980s found out to their cost.

House price inflation does not proceed smoothly. It is a cyclical process and the British home ownership market is no exception to this generalisation. A series of booms, of greater or lesser duration and magnitude, are generally followed by periods of relative stability, downturns or, worst of all, a sharp

Table 2.1   The relationship between house prices and earnings

	Average house price		Average earnings		House price earnings ratio	Change in retail prices %	Change in real house prices %
	£	% Change YoY	£	% Change YoY			
1956	2,230		693		3.22	5.1	
1957	2,280	2.2	727	4.9	3.14	3.7	−1.4
1958	2,360	3.5	752	3.4	3.14	3.0	0.4
1959	2,360	0.0	789	4.9	2.99	0.5	−0.5
1960	2,480	5.1	844	7.1	2.94	0.9	4.2
1961	2,710	9.3	891	5.5	3.04	3.6	5.5
1962	2,890	6.6	917	2.9	3.15	4.2	2.4
1963	3,100	7.3	961	4.8	3.23	1.9	5.2
1964	3,390	9.4	1,034	7.7	3.28	3.3	5.8
1965	3,740	10.3	1,108	7.1	3.38	4.7	5.3
1966	4,040	8.0	1,180	6.6	3.42	4.0	3.9
1967	4,270	5.7	1,223	3.6	3.49	2.5	3.1
1968	4,650	8.9	1,319	7.8	3.53	4.6	4.1
1969	4,850	4.3	1,422	7.8	3.41	5.5	−1.2
1970	5,190	7.0	1,586	11.5	3.27	6.3	0.7
1971	6,130	18.1	1,742	9.8	3.52	9.4	7.9
1972	8,420	37.4	1,953	12.1	4.31	7.1	28.3
1973	11,120	32.1	2,237	14.5	4.97	9.2	20.9
1974	11,300	1.6	2,644	18.2	4.27	16.0	−12.4
1975	12,119	7.2	3,302	24.9	3.67	24.2	−13.6
1976	12,999	7.3	3,802	15.2	3.42	16.6	−8.0
1977	13,922	7.1	4,147	9.1	3.36	15.9	−7.6
1978	16,297	17.1	4,723	13.9	3.45	8.2	8.2
1979	21,047	29.1	5,473	15.9	3.85	13.4	13.9
1980	24,307	15.5	6,592	20.5	3.69	18.0	−2.1
1981	24,810	2.1	7,443	12.9	3.33	11.9	−8.8
1982	25,553	3.0	8,139	9.4	3.14	8.6	−5.2
1983	28,592	11.9	8,825	8.4	3.24	4.6	7.0
1984	30,811	7.8	9,362	6.1	3.29	5.0	2.7
1985	33,187	7.7	10,141	8.3	3.27	6.1	1.6
1986	38,121	14.9	10,948	8.0	3.48	3.4	11.1
1987	44,220	16.0	11,799	7.8	3.75	4.1	11.4
1988	54,280	22.7	12,829	8.7	4.23	4.9	17.0
1989	60,763	11.9	13,993	9.1	4.34	7.8	3.8
1990	65,059	7.1	15,358	9.8	4.24	9.5	−2.2
1991	65,593	0.8	16,588	8.0	3.95	5.9	−4.8
1992	63,638	−3.0	17,596	6.1	3.62	3.7	−6.5
1993	67,029	5.3	18,219	3.5	3.68	1.6	3.7
1994	68,892	2.8	18,949	4.0	3.64	2.5	0.3
1995	68,551	−0.5	19,584	3.4	3.50	3.4	−3.8

Source: Council for Mortgage Lenders, *Housing Finance* No 31, August 1996

slump when both sales volumes and prices fall sharply (Fleming and Nellis, 1990; Halifax BS, 1992). Such booms and slumps are not uncommon in other countries and they are common to most asset markets where, when prices begin to rise sharply, and buyers fear they may miss the boat, a period of frenzied speculative activity can set in, with buyers bidding up prices to levels which later prove to be unsustainable (Hendry, 1984; Case, 1986; Case and Shiller, 1988). When the inflationary bubble bursts (which can happen for a variety of reasons, but usually rests on the fact that prices have risen so far above incomes that buyers are priced out of the market and houses become unaffordable), prices can fall sharply as buyers disappear, and sellers either try to sell at the best price they can get or simply sit on their assets and wait for better times (Muellbauer, 1990c).

The housing market differs from many other speculative asset markets in that housing provides an important use value as well as an exchange value. Houses first and foremost provide a roof over one's head, as well as a store of value and a potential source of capital gains. Also, most owners own only one property, and they are not able to take speculative positions (except in the sense that they may decide to move up to a more expensive property in the expectation of future prices inflation). Consequently, the housing market arguably has a greater degree of stability than many other asset markets, in that most people (speculators apart) live in their asset. Nonetheless, there are always cyclical trends in prices and sales volumes as the housing market heats up or cools down after a boom. Australia and New Zealand have had several such slumps since 1945 (Badcock, 1989; Dupuis and Thorns, 1997; Abelson, 1997; Thorns, 1989; Dupuis, 1992). The Netherlands experienced a major boom from 1974 to 1977 followed by a deep slump from 1979 to 1982 and then by stagnation until the early 1990s. Dieleman (1992) reports that the price of an average house in the Netherlands in 1970 was Dfl 49,100. By 1978 average prices had risen to Dfl 166,000, a real gain of Dfl 38,000. Between 1978 and 1982 the average price fell to Dfl 129,000: a nominal loss of Dfl 37,000 and a real loss of Dfl 67,000. Not surprisingly, the Dutch home ownership market took many years to recover from this shock. Similar cycles have been observed in Finland (Ruonavaara, 1993), Sweden (Turner, 1993) and the USA (Tucillo 1980; Grebler and Mittelbach, 1979). In Finland, which experienced a dramatic boom and bust in the late 1980s and early 1990s, prices rose by 40% during 1988, but from 1989 to the end of 1992 prices fell by 35% and the price of flats halved between 1989 and 1992 (Ruonavaara, 1993). Harris (1986) has pointed to the existence of a similar, if less marked, cycle in Canada, particularly in the cities of Montreal, Toronto and Vancouver, and property prices have been extremely volatile in Hong Kong and Japan (Lui, 1995: Hirayama and Hayakawa, 1995). House prices seem to be cyclical in most western countries, though much less so in France and Germany than in other western European countries. In an analysis of house price volatility in western Europe 1985–93, Stephens (1995) points out that the Scandanivian countries, Sweden, Norway, Finland and Denmark, experienced the largest nominal house price falls, with Britain coming in fifth place.

21

The British market has also been strongly cyclical but, with the exception of the early 1950s, prices had never fallen in cash terms during the downturns until the slump of the early 1990s, although they fell in real, that is to say in inflation-adjusted, terms. Consequently, Britain never experienced a really deep slump until the early 1990s (Breedon and Joyce, 1992; Cutler, 1995). Previous postwar downturns were more akin to breathing spaces between booms, during which prices marked time in cash terms. The 1990s slump marked a radical departure from previous downturns in that it was the first one where cash prices actually fell sharply rather than prices simply falling in real terms. It was this feature of the slump more than anything else which generated severe problems, as in previous downturns most sellers had been able to sell for what they had paid in cash terms, even though they may well have taken a loss in real terms. Given the conjunction of the housing market slump with the recession of the early 1990s, and the sharp rise in unemployment, people who had bought in the mid to late 1980s, at or near the peak of the boom, and subsequently lost their jobs or split up with partners, found that if they defaulted on their mortgage payments or were forced to sell they faced negative equity and a substantial loss. This in turn lead to a massive rise in mortgage arrears and to building society and bank repossessions (Breedon and Joyce, 1992). A significant part of the problem stemmed from the fact that deregulation of mortgage lending in the early 1980s, combined with greater competition between mortgage lenders to expand or retain market share, led to the growth of generous mortgage advances based on high income multiples. This helped to push up prices to unsustainable levels in relation to incomes and caused repayment problems when mortgage rates and unemployment rose in the late 1980s and the boom burst. This chapter goes on to examine the nature, causes and consequences of the 1980s boom and bust and put it in its historical context.

The Evolution of the UK Home Ownership Market from the 1950s to 1989

In the 1950s and 1960s the home ownership market in Britain was relatively stable. In 1956, the first year for which Building Societies Association statistics are available, the national average house price was £2,230. House prices rose slowly through the second half of the 1950s to an average £2,480 in 1960; by 1969, on the eve of the first major house price boom, the average price had doubled to just under £5,000 – a rise of 100% in ten years or an average of 8% per year (Building Societies Association, 1990). The annual rate of house price inflation varied, but never exceeded a peak of 10% and the average annual rate of house price inflation in the peak years of 1964 to 1966 was 9% (Figure 2.2). The house price/income ratio, which measures the relation between national average house prices and national average male income, fell during the late 1950s from 3.19 in 1956 to 2.92 in 1960. It then rose to a peak of 3.5 in 1968, falling back slightly in 1969 and 1970 to 3.2 (Figure 2.3).

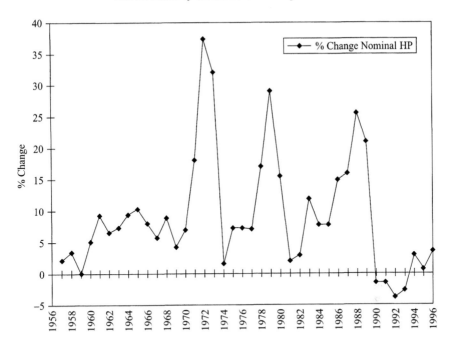

Figure 2.2 Annual percentage change in nominal UK house prices, 1956–96
Source: Council for Mortgage Lenders (1995, 1997)

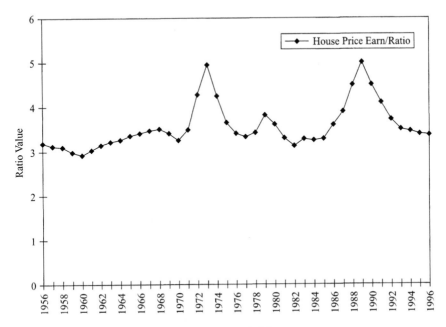

Figure 2.3 UK house price/earnings ratio, 1956–96
Source: Council for Mortgage Lenders (1995, 1997)

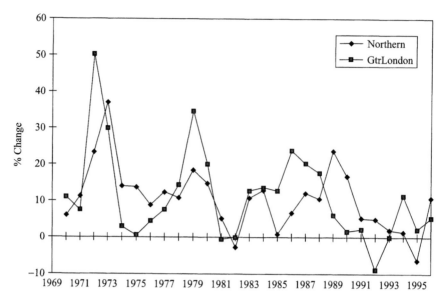

Figure 2.4 Percentage annual change in house price inflation, London and the North 1970–96

Source: Council for Mortgage Lenders (1995, 1997)

The 1970s saw a dramatic break with the preceding fifteen years of slow but steady house price inflation. The year 1971 saw the beginning of the first of the three major house price booms of the 1970s and 1980s. In 1971 national average prices rose by 18% (Building Societies Association, 1990) and in 1972 and 1973 they rose by 37% and 32% respectively. The national average house price stood at £11,100 in 1973: almost double the £5,200 of 1970. It is difficult to say precisely what caused the first major postwar boom, but evidence points strongly to two factors. The first was a demographic shock. The postwar baby boom of 1945–48 were beginning to enter the labour and housing market in large numbers in 1971. The 25–29 age group (the peak home-buying age) grew sharply in size from 1970 to 1973 (Breedon and Joyce, 1992; Cutler, 1995); see also Milne and Clark 1990; Lee and Robinson, 1989a; 1990; Ermisch, 1990). The private rented market was already contracting sharply (Hamnett and Randolph, 1987) and many members of this age group were looking to buy. With a static supply of housing, prices began to rise rapidly, assisted by Barber's early 1970s Conservative credit boom which made borrowing far easier than hitherto. Until 1974 interest on all loans could be offset against income tax, including borrowing for second homes and speculative purposes. Not surprisingly, a wave of mortgage borrowing flooded into the market, helping to push prices higher as buyers competed (Muellbauer, 1990c).

The boom began in London and the South East and subsequently spread to other regions, though with a slight lag. This can be seen in Figure 2.4, which shows annual rates of house price inflation in London and the North. At the peak of the boom in 1972–73 the gap between average house prices in London and the South East on the one hand and the Midlands/North on the other reached its peak. At this point the average property in London cost almost twice as much as a similar property in the North. This trend has also characterised the subsequent two booms, leading to a longstanding debate over the alleged widening of the North–South house price gap. But, as will be shown later, the regional house price gap is essentially cyclical, widening rapidly at the start of a boom as house prices increase most rapidly in London and the South East, and subsequently falling as house prices stabilise in the South East and continue rising elsewhere.

The house price/earnings ratio rose rapidly from a low of 3.25 in 1970 to a peak of 4.95 in 1973. Mortgage payments as a proportion of average first-time buyer income also rose from 11.6% in 1971 to a peak of 15% in 1976. Such high ratios are characteristic of the peak of house price booms but are unsustainable in anything but the very short term, and they inevitably prompt a correction to bring house prices back broadly in line with incomes. Some trigger is required, however, and as in subsequent booms this was provided by government policy. The house price boom of the early 1970s was brought to an abrupt halt in November 1973 by the overnight increase in interest rates of five percentage points in a desperate bid to bring inflation under control. The result was wholly predictable. The market in the overheated South East collapsed almost overnight. Turnover fell sharply, and building society net advances fell from £2 billion in 1973 to £1.5 billion in 1974. House prices did not fall in nominal terms in any region, but the national average rate of house price inflation fell to 1.6% in 1974 and to 7% in each of the three following years.

The real impact of the slump only appears when we take overall inflation into account. The retail price index rose by 16% in 1974, 24% in 1975, and 16% in 1976 and 1977: the highest level of inflation in the postwar period. Real house prices fell by 12% in 1974, 14% in 1975 and 8% in each of the two subsequent years. Anyone who bought a house in London and the South East in 1973 (myself included) would have seen its real value fall by a third by 1977 (Figures 2.5 and 2.6). Nominal prices did not fall because the very rapid general inflation underpinned nominal prices. These are national average figures, however. Nominal house prices rose by much less in London and the South East than they did in the Midlands and the North during the mid 1970s. Thus the real price falls incurred in the South East were far greater. Essentially, house prices rose first and most rapidly in London and the South East. House price inflation in London and the South East was very low during the mid 1970s (2–3%) while it continued at 7–8% in the Midlands and the North and 10–15% in Scotland. Thus, the North–South gap was at its widest during the early years of the boom and narrowed during the mid 1970s, and this pattern has been characteristic of subsequent booms and downturns. A slump in the South East often provides a catching up period for prices in the Midlands and the North (Figure 2.7) and the house price gap narrows, although most observers fail to notice this as press attention is focused on the periods when the gap widens and dramatic headlines beckon.

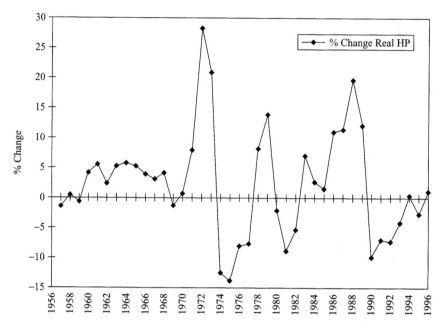

Figure 2.5 Annual percentage change in real UK house prices, 1956–96
Source: Council for Mortgage Lenders (1995, 1997)

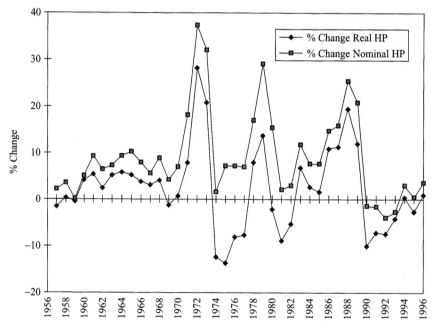

Figure 2.6 Annual percentage change in nominal and real UK house prices, 1956–96
Source: Council for Mortgage Lenders (1995, 1997)

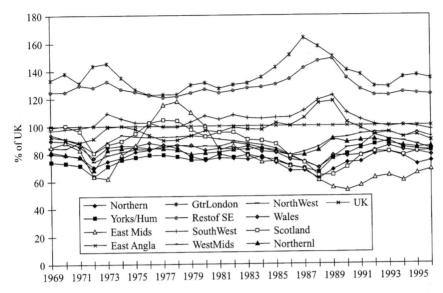

Figure 2.7 Regional average house prices relative to the UK average, 1969–96

Source: Council for Mortgage Lenders (1995, 1997)

New building of private housing fell back sharply from an average of 192,000 completions a year in 1971–73 to 141,000 in 1974 and 151,000 in 1975 and 1976. The volume of sales also slumped from 1.4 million in 1973 to under 1 million in 1974; the number of new mortgage loans fell from a peak of 750,000 in 1973 to just 550,000 in 1974 and mortgage advances as a proportion of price paid fell from 82% in 1972 to 72% in 1974 (Figure 2.8). The house price/earnings ratio rapidly fell back from its peak of 4.95 in 1973 to 4.25 in 1974 and to 3.34 in 1977 as house prices and incomes were restored to equilibrium, and mortgage payments as a proportion of average earnings also fell back from their peak of 15% in 1973 and 1974 to a low of 11.6% in 1978. Conditions were now ripe for the next boom.

Each cycle has been characterised by essentially the same pattern. At the start of each upturn, real incomes tend to rise faster than house prices and the house price/income ratio is at or below its long-term historic norm of 3.5. As the volume of sales rises, any overhang of unsold property on the market begins to dry up and (as housing supply is largely fixed), prices begin to rise quite sharply. The house price/income ratio rises and, as the cycle nears its peak, house prices rise ahead of incomes to the point where the ratio becomes unsustainably high and prices stabilise and fall in real terms. Subsequently, as incomes continue to rise, the house price/earnings ratio falls back until the conditions are in place for a new boom. Figure 2.9 shows the relationship between incomes and house prices.

The second boom, which began in 1978, followed a similar, though less marked cycle. Real earnings growth, which had been negative from 1974–77 (as a result of the very high rate of inflation) rose very sharply to 7% in 1978 and

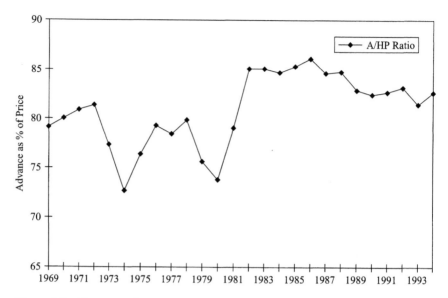

Figure 2.8 Mortgage advance as percentage of price paid, first time buyers, 1969–94
Source: Council for Mortgage Lenders (1995, 1997)

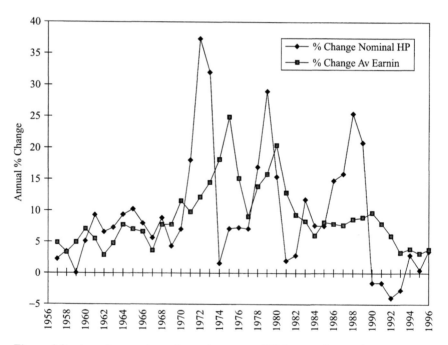

Figure 2.9 Annual percentage change in average UK house prices and average earnings, 1956–96
Source: Council for Mortgage Lenders (1995, 1997)

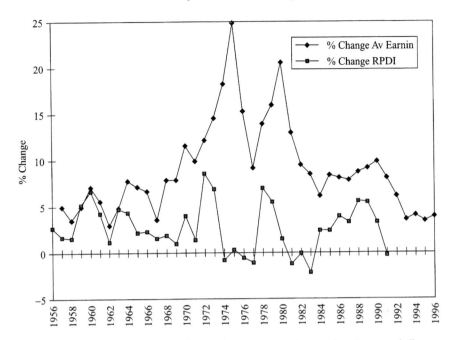

Figure 2.10 Annual percentage changes in average earnings and real personal disposable income, 1956–96

Source: Council for Mortgage Lenders (1995, 1997)

5.4% in 1979 (Figure 2.10). Nominal house price inflation also rose sharply, from an average of just over 7% in 1975–77 to 17% in 1978, 29% in 1979 and 15.5% in 1980. The national price/income ratio (which had fallen to 3.34 in 1977) rose sharply to a peak of 3.82 in 1978, and initial mortgage payments as a proportion of income rose from a low of 11% in 1978 to a high of 17.4% in 1980. Once again, London and the South East led the boom, with price rises of 35% and 30% in 1978/9 compared to a rise of just over 20% in Scotland and the North (Figure 2.4).

As with the early 1970s boom, these ratios were unsustainable, and with the trigger of rapid rises in interest rates (to bring borrowing back under control) from a low of 8.5% in January 1978 to 15% by November 1979, the boom quickly subsided. During this period the national average house price rose from £13,000 in 1976 to £24,300 in 1980. The national average rate of house price inflation fell back to 15.5% in 1980 and just 2.1% and 3% in 1981 and 1982. Once again, real rates of house price inflation were negative in 1980–82 (–9% in 1981), but nominal house prices only fell very slightly in London while continuing to rise slowly in the Midlands and the North. Initial first-time buyer mortgage repayments as a proportion of average income fell back from a peak of 18.9% in 1980 to 14.5% in 1983 (Figure 2.11) and the house price/earnings ratio fell back to a low of 3.13 in 1982, setting the scene for the third and most destructive boom.

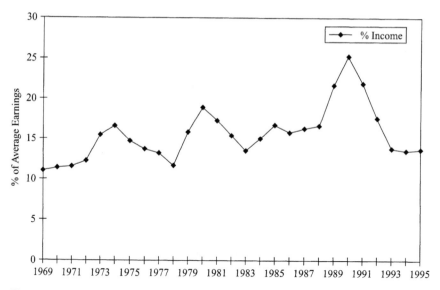

Figure 2.11 Initial mortgage repayment as percentage of average income, first time buyers, 1969–95

Source: Council for Mortgage Lenders (1995, 1997)

Real incomes rose steadily from 1983 to 1987 by about 3% a year on average and mortgage rates fell from 14% in 1985 to a low of 9.5% in May 1988. But more importantly, the early 1980s saw the entry of the banks into the home mortgage market, the rapid growth of competition and the relaxation of traditional mortgage to income and advance to price ratios. The expansion of mortgage lending from £7.3 billion in 1980 to £14.1 billion in 1982 coincides with the entry of the banks into the mortgage market. They cut back new lending from 1983 to 1986, but bank lending jumped sharply in 1987 once again. In addition a variety of new central mortgage lending institutions entered the market with a big impact from 1986 onwards. The result was a surge of lending which peaked in 1988 at a remarkable £40.2 billion (a 430% increase on 1981). The massive surge in mortgage lending from 1981 onwards, which was even greater from 1985 to 1988 is shown in Figure 2.12. There was also a big jump in mortgage advances as a percentage of price paid from a low of 74% in 1980 to 85% in 1982 (Figure 2.8). The number of mortgage loans rose sharply from the early 1980s when the banks first entered the home mortgage market, doubling from 750,000 in 1980 to over 1.5 million in 1986. The number of building society loans rose more gradually: from 675,000 in 1980 to 1.232 million in 1988 (Council for Mortgage Lenders, 1992). This was linked with a gradual rise in the number of residential transactions from a low of 1.27 million in 1980 to a peak of 2.0 million in 1988 (Figure 2.13).

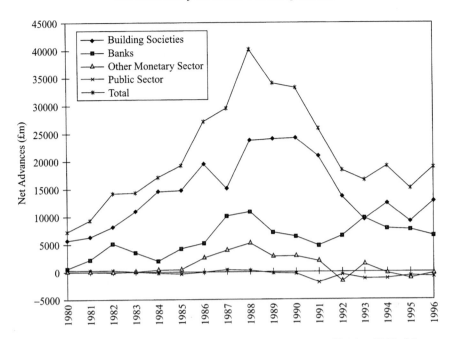

Figure 2.12 Loans for house purchase (net advances) by type of lender, 1980–96
Source: Council for Mortgage Lenders (1995, 1997)

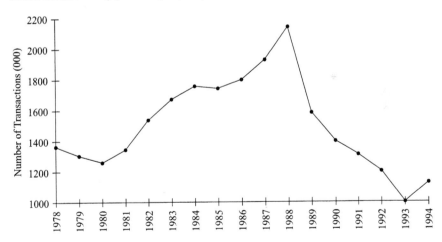

Figure 2.13 Number of residential property transactions in England and Wales, 1978–94
Source: Inland Revenue Statistics (various years)

Muellbauer suggests that a shock is needed for every boom to begin, and the shock *par excellence* of the 1980s was financial liberalisation. Mortgage lending increased rapidly as the banks entered the market, and lenders competed for market share by offering ever more generous mortgage to income and mortgage advance to price ratios (which reached 100% at the peak of the boom). The boom was also underpinned by a sharp increase in the size of the 20–29 age group, which grew by 1.3 million from its low point in 1978 to an all-time high in 1990 as the baby boom of the 1950s and 1960s entered the housing market. Cutler (1995) also suggests that the rate of household formation increased as first-time buyers brought forward entry into owner occupation as prices rose, but this was probably only true of the later stages of the boom (Holmans, 1995).

The third house price boom was the longest of the three booms of the 1970s and 1980s. It began in 1983 when national average nominal prices rose by 12%, but they then subsided in 1984 and 1985 to 7–8%. The boom properly got under way in 1986 when national average prices rose 15%, followed by rises of 16% and 23% in 1987 and 1988. Unlike the previous two booms, the third boom was initially concentrated in London and the South from 1985 to 1988. The boom only spread to the Midlands and the North in 1988 and 1989, while it had reached its peak in the southern regions in 1988 and begun to subside rapidly (Figure 2.4). This boom was so regionally lagged as to be almost counter cyclical. Average mix adjusted house prices increased by 15% in the South East, 17% in East Anglia and 9% in London in 1989, compared to increases of 30–40% in the Midlands and the northern regions. House price inflation diffused outwards from London and the South East like the ripples in a pond, reaching a peak in successively later time periods further from London as the wave passed (Coombes and Raybould, 1991).

Initial first-time buyer mortgage repayments as a proportion of average income rose from 17% in 1987 to a high of 26% in 1990. The house price earnings ratio rose to a peak of 4.43 in 1989 (and over 5 in London and the South East). These ratios were unsustainable in anything but the short run as buyers were massively overstretched, and the house price bubble burst in late 1988. Once again, the third slump was triggered by government policy. In the April 1988 Budget it was announced that mortgage interest tax relief was to be limited to one mortgage per dwelling as from 1 August. This was designed to curtail growth in the number of unrelated individuals taking out mortgages to buy a property in the South East. This announcement had the effect of bringing forward many potential purchases to beat the deadline. In addition, the Government announced sharp rises in interest rates in August 1988 in order to try to control credit growth and consumer spending. These were running out of control partly, as we shall see in Chapter 5, because of home owners' equity withdrawal from the housing market. As a result, mortgage interest rose sharply from 9.75% in May 1988 to 12.75% in September 1988, and a peak of 15.5% in February 1990: an increase of over 50% in under two years.

The impact of such a rise on the budgets of recent first-time buyers who had stretched themselves to buy in the late 1980s when house price/income ratios were at their peak is easy to appreciate. Some were already on the margin with large mortgage repayments they were only just able to afford. The increase in interest rates proved the last straw and many were pushed over the brink into mortgage debt. Their difficulties were intensified by the onset of the recession in 1991–92 which saw unemployment rise from 1.8 million in 1990 to over 3 million in 1992. This was, in turn, partly brought about by the sharp rise in interest rates which tipped many businesses into receivership. When mortgage interest rates rose and house prices fell, many recent buyers who had bought on high mortgage to price ratios had no safety margin. Overgeared and over-committed, they fell victim to the inevitable correction in prices and demand.

The 1989–95 Slump

The boom collapsed like a pricked balloon. The number of sales fell from a peak of 2 million in 1988 to 1.47 million in 1989, 1.28 million in 1990 and 1.03 million in 1992. The initial fall in transactions was almost entirely concentrated in the four southern regions, where the boom had previously been marked. Between 1988 and 1989 the number fell by 36% in Greater London, 42% in the South East, 41% in the South West, 48% in East Anglia and 37% in the East Midlands. In the West Midlands and the northern regions, by contrast, transactions remained static or even rose slightly (Figure 2.14, Table 2.2). Between 1989 and 1990 sales volume stabilised in the southern regions with falls of between 1% and 5%, and transactions rose in East Anglia by 11%. By then, the

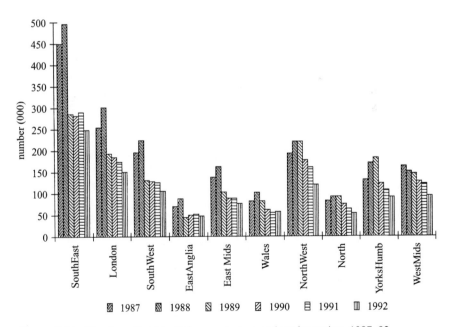

Figure 2.14 Number of residential property transactions by region, 1987–92

Source: Inland Revenue Statistics (various years)

Table 2.2 Residential transactions by region, 1986–91

	Number of transactions (000)						% change		
	1986	1987	1988	1989	1990	1991	1988–89	1989–90	1990–91
E Anglia	62	70	86	45	50	52	−47.7	11.1	4.0
SE	419	448	497	287	283	291	−42.3	−1.4	2.8
SW	173	194	222	130	127	125	−41.1	−2.3	−1.6
London	240	255	302	192	183	172	−36.4	−4.7	−6.0
E Mid.	114	136	159	101	87	87	−36.5	−13.9	0.0
Wales	73	81	100	82	61	55	−18.0	−25.6	−9.8
W Mid.	144	159	147	143	125	119	−2.7	−12.6	−4.8
NW	170	191	220	219	174	158	−0.5	−20.5	−9.2
North	73	81	90	90	73	61	0.0	−18.9	−16.4
Yorks/H	131	129	168	179	121	105	6.5	−32.4	−13.2
England/Wales	1,599	1,744	1,991	1,468	1,284	1,225	−26.3	−12.5	−4.6

Source: IRS Residential Transactions by Region, 1991, 1992, 1993

decline had begun to spread to the northern regions. Not surprisingly, there was also a sharp fall in the volume of mortgage lending. The number of loans fell from 1.65 million in 1988 to 1.06 million in 1991 and to 900,000 in 1992.

The 1990s slump differed from previous market downturns in a number of key respects. First, although real house prices fell steeply, the size of the fall was far greater than in previous downturns. Second, there was a damaging and unprecedented fall in nominal house prices which led to a growing number of owners with negative equity estimated at 2 million at the peak (Chapter 4) and a rise in the number of mortgage arrears and repossessions. This led to a wider crisis of confidence in the future of house prices and the ownership market in general which became more entrenched as the slump continued. This loss of confidence combined with the long duration of the slump and the sharp rise in repossessions was quite unprecedented. As the Halifax Building Society (1992: 4) commented: "The most significant difference between the events of the 1988–92 down-turn compared with those of 1974–76 or 1979–82 is that nominal house prices have fallen while mortgage debts have remained constant in money terms and interest charges have stayed positive. People are thus less wealthy." Riley (1992) noted in the *Financial Times* that: "Although house prices have been known in the past to fall in real terms the nominal decline of the past three or four years is quite unprecedented in the living memory of homeowners".

Riley was entirely correct in his assessment, but the standard DoE/BSA 5% sample of mix unadjusted house price completion data fail to show the falls. On the contrary, this source shows national average house prices rising from £54,850 in 1989 to £62,455 in 1992, falling slightly to £60,800 in 1993, before recovering to £65,644 in 1995. So, too, prices in London apparently rose from an average £82,380 in 1989 to £85,740 in 1991 before falling back to £78,250 in

Table 2.3 Comparison of mix-adjusted and unadjusted prices in London, 1989–93

	1989 £	1991 £	1993 £	1989–93 loss (£)	% change
DoE/BSA mix-unadjusted	82,383	85,742	78,399	−3,984	−4.8
DoE/BSA mix-adjusted	98,700	93,100	82,200	−16,500	−16.7
Nationwide mix-adjusted	97,667	77,901	66,948	−30,719	−31.5

Source:

Table 2.4 Regional house price changes in the UK, 1983–97

	Q1 1983 £	Q3 1989 £	Q1 1993 £	% Rise 83–89	% Fall 89–93	Q3 1997 £	% Rise 93–97
London	34,426	97,667	66,948	183.7	−31.5	95,481	+42.6
OMA	35,744	99,979	69,297	179.7	−30.6	89,152	+28.6
OSE	29,314	83,901	54,455	186.2	−35.1	70,531	+29.5
E Anglia	26,216	78,846	50,609	200.7	−35.8	62,683	+23.9
SW	27,009	73,112	53,954	170.7	−26.2	64,270	+19.1
E Mid.	22,521	57,840	45,341	156.8	−21.6	54,796	+20.8
Yorks/H	25,839	64,936	49,095	151.3	−24.3	53,382	+8.7
W Mid.	23,651	61,290	52,141	159.1	−14.9	60,617	+16.3
Wales	23,849	54,188	48,422	127.2	−10.6	51,524	+6.4
NW	22,666	51,467	51,673	127.1	+0.4	55,053	+6.5
North	20,955	43,195	45,384	106.1	+5.1	47,048	+3.7
Scotland	26,375	47,181	52,469	78.9	+11.2	57,101	+8.8
N Ireland	23,293	28,539	34,574	22.5	+21.1	51,962	+50.3
All UK	26,307	62,244	50,128	136.6	−19.5	61,830	+23.3

Source: Nationwide Building Society 1995: *House Price Data Supplement*, Historical House Price Series

1992. East Anglia showed a clearer fall from £64,600 in 1989 to £56,770 in 1993. These unadjusted figures are contrary to the Nationwide and Halifax mix-adjusted house price series and what seems to have happened is that the slump led to a change in the mix of sales, with fewer low-priced flats and a higher proportion of the more expensive houses. This meant that the standard DoE/BSA 5% mix unadjusted price data did not show price falls in any region until 1991 and then only very small ones in East Anglia and the South East. The DoE/BSA mix-adjusted data provide a more realistic picture with UK average prices falling from £70,400 in 1989 to £64,300 in 1993, from £75,300 in 1989 to £58,700 in 1993 in East Anglia and from £98,700 to £82,200 in London (Table 2.3). Leaving this problem aside, Nationwide Building Society (1995) figures show national average house prices for all properties rose from £26,307 in Q1 1983 to £62,244 in Q3 1989 before falling to £50,128 in Q1 1993: a 20% decline nationally (Table 2.4).

The decline in prices was most dramatic in the South East of England. The Nationwide house price data show that average prices in London rose from £34,426 in Q1 1983 to a peak of £97,667 in Q3 1989 (almost a threefold increase) before falling to £66,948 in Q1 1993. This is a fall of 31% in cash terms. Prices fell by a similar proportion in the Outer Metropolitan Area (OMA), and by 35% in the Outer South East (OSE) and East Anglia (Table 2.4). In real terms, the fall in prices has been even more marked. In the rest of the country house prices rose by considerably less during the 1980s, and the fall in prices in the 1990s was consequently much less than in southern Britain.

The sharp falls in house prices in southern England in the 1990s have meant that the regional house price gap contracted sharply but this has not been picked up by most commentators as it is an example of the "small earthquake, nobody killed" story. The Nationwide Building Society (1997) noted that the ratio of house prices in London and South East to the rest of England and Wales fell from a peak of almost 2.0 in early 1988 to a low of 1.3 in early 1993 and is climbing upwards once again. The regional house price gap is strongly cyclical.

Arrears and Repossessions as Indicators of the Severity of the Slump

Falling prices and transaction volumes are good indicators of a change in market activity. But the severity of the slump was shown by the dramatic rise in mortgage arrears and repossessions. Under normal market conditions a borrower who gets into payment arrears, perhaps because of losing their job, can reach an accommodation with the lender about repayments and can sell the property if necessary. But, during the slump, many borrowers found that the value of their house was less than the outstanding mortgage. The number of repossessions and arrears had been rising since the early 1980s, as a result of growing unemployment, with a slight decline in late 1988 and the first half of 1989 during the consumer boom, but the big increase was in 1990 onwards as the full effects of the recession worked their way through. The number of repossessions rose from 14,580 in 1988 to 75,540 in 1991: an increase of 400%, gradually falling to 49,300 in 1994 and 1995 and 42,560 in 1996. In the period 1990–96 the number of repossessions totalled 387,690, almost 4% of the total number of mortgages. The number of mortgages 6–12 months in arrears rose from a low of 21,540 in 1981 to a peak of 205,010 in 1992: an almost tenfold increase, falling back to 101,960 in 1996, and the number of mortgages 12 months or more in arrears rose from 5,540 in 1982 to a peak 151,810 in 1993: a 26-fold increase, falling back to 67,020 in 1996 (Figure 2.15). As Pannell and Abisogun (1992: 17) put it:

> The severity of the downturn in the UK housing market has been aggravated by the number of possessions, which have acted both directly in pressuring house price levels and indirectly through their impact on

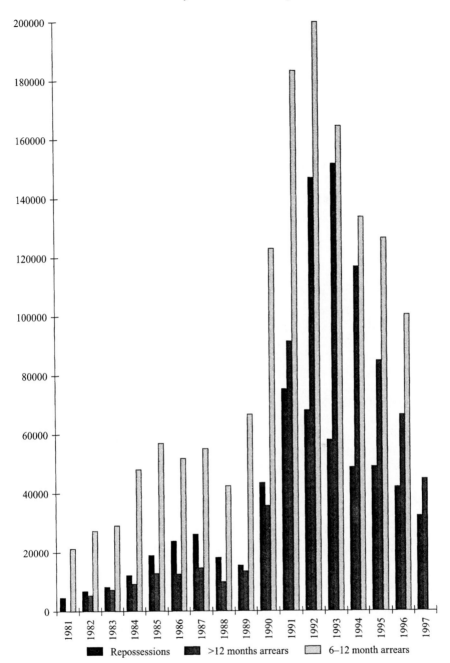

Figure 2.15 Mortgage arrears and repossessions, 1981–97

Source: Council for Mortgage Lenders (1997)

confidence. It is also argued that negative equity has cast a pall over the housing market, with many households financially unable or strongly disinclined to move house. To the extent that this factor is at work, it could, perhaps, be seen as part of a wider-reaching change in the housing market as a result of a re-evaluation of the investment appeal of owner-occupation, although the evidence of a number of recent market research exercises is that the underlying demand for owner-occupation remains strong.

This view is strongly supported by Breedon and Joyce (1992), who suggest that the rise in possessions to the point where they comprised 5% of turnover in 1992 may have had a significant depressing effect on house prices in that it reduced demand relative to supply. They suggest that a more lenient stance on repossessions could help house prices recover. As Stephens (1995: 561) points out: "Debt deflation occurs because the collective impact of borrowers selling houses to avoid negative equity and lenders repossessing houses has a downward pressure on house prices so pushing more borrowers into negative equity and causing more of the lenders' security to disappear." He argues that government interventions in the housing market in 1991–93 were directed at reducing repossessions by gaining the collective agreement of the lenders to reduce repossessions in return for reforms of the method of payment of social security support to home owners with a mortgage and then keeping the houses off the market by financing housing associations to buy them.

One of the most important aspects of the slump is its duration. In previous downturns the market was depressed for a maximum of three to four years (1974–77 and 1981–83). The 1990s slump lasted from 1989 to 1995 and increased in severity during the early 1990s as prices continued to fall and then bounced along the bottom. Consequently, even as late as 1995 prices were still falling nationally. It is therefore not surprising that housing market activity continued at a depressed level. In some respects it is remarkable that transactions remained at around 1.2 million per annum during the 1990s as until the end of 1993 (in the South) buying a house was often a passport to losing money.

The Impact of the Slump on the Mortgage Lending Institutions

The surge in mortgage arrears and repossessions did not just hit borrowers: it also had an impact on the mortgage lenders, who faced a significant rise in bad debts. The five largest mortgage lenders, the Halifax, Nationwide, Woolwich and Leeds Permanent Building Societies and the Abbey National which converted from a building society to a bank in 1989, experienced a dramatic worsening in provisions for bad debts and write-offs between 1990 and 1993. In 1990 their combined provisions for bad debts on residential property totalled £164 million but by 1993 this had risen to £1,278 million: an increase of 679%. Their collective write-off on irrecoverable loans rose from £18 million in 1990 to £563 million: an increase of 2,943%. Taking 1992 and 1993 together, their provisions totalled £2.376 billion and their write-offs £1.03 billion (Stephens, 1996). Their profits fell by only 4% from 1990 to 1993, however, which reflects the fact that they were able to offset losses against mortgage indemnity policies taken out with insurance companies which gave protection to lenders for the first 25% of any loss arising from loan default and repossession. The cost of these losses for insurance companies rose from £48 million in 1990 to a peak of £1.37 billion in 1991. Stephens (1996: 340) notes that the new centralised lenders such as the National Home Loans Corporation and Mortgage Express generally fared worse than the building societies in that they:

> faced higher levels of loan delinquency than the traditional lenders be-cause they entered the market when lending criteria were generally loose and had no stock of older, safer, mortgage assets . . . their lending criteria were even looser than that of the traditional lenders, since they were aiming to gain market-share quickly. And of course, they entered the market as it neared its peak, so their borrowers were particularly vulner-able to rising interest rates and falling house prices.

As a consequence, many centralised lenders ceased to make new loans and others withdrew from the market altogether. In some cases, they are still picking up the pieces after their ill-timed foray into the mortgage market. Stephens notes that the depth and duration of the recession, combined with the rapid growth in negative equity, mortgage arrears and repossessions, also prompted calls for government intervention to rescue home owners and the housing market. And as the housing market slump was also seen to be a causal factor in the general economic recession (Chapter 8) there were calls for the Government to intervene to kickstart the economy. But these went largely unheeded.

The Conservatives were not sympathetic to a bust which was a product of the previous boom funded by generous mortgage lending and government interven-tion was limited despite the potential electoral problems caused by negative equity and repossessions and the Government's encouragement of mass home ownership. Nor did the mortgage lenders seem disposed to take dramatic action. By the time some of the proposed measures were in place, which allowed owners with negative equity to move house, the slump was almost over (Stephens, 1993a; Clapham, 1996). In general, little of significance was achieved by lenders and policy makers although the building societies did ease up on repossessions from 1992 as the flood of repossessed properties on the market was depressing property values further. Thus while long-term arrears rose by 60% in 1992, repossessions fell by 9%. In return, the Government agreed that the social secur-ity payments for borrowers' mortgage interest payments would be paid directly to lenders rather than to borrowers (Figure 2.16).

The Government also temporarily raised the threshold for stamp duty on new purchases from £30,000 to £250,000 for eight months in 1992 in an attempt to jump-start the market, and in 1993 it doubled the threshold to £60,000. This had

Figure 2.16 "You have been keeping up the mortgage repayments, haven't you?"

Source: The Independent, Thursday 19 December 1991

little effect on demand, not least because the Government cut the rate of mortgage interest tax relief to 15% in order to reduce the rapidly growing cost of the subsidy, which had risen from £3.58 billion in 1983 (at 1993 prices) to £8.14 billion in 1991. It fell to £5.2 billion by 1993. The Government also announced an increase of some £600 million to housing association grants in 1993 to allow them to purchase properties which had been repossessed and developers' unsold houses in order to prevent them further depressing the market; Stephens (1996) questions the success of this measure, despite the fact that more than 18,000 properties were taken off the market in England.

The Estate Agency Débâcle

There was another disastrous consequence of the financial deregulation and liberalisation: the huge losses incurred by a number of financial institutions as a result of their ill-judged decision to expand into estate agency in the late 1980s. The rationale for the expansion was that estate agents had considerable potential for originating loans and selling a wide variety of financial services at the point of house purchase through their extensive network of offices. Also estate agency was seen to be a profitable business in its own right, and control of what was a fragmented small-business sector offered a prospect of higher margins. This view had considerable logic. The principal problem lay in the timing and the financing of the expansion. The first institution to expand was Lloyds Bank, which set up Black Horse Agencies in 1982 following their acquisition of a medium-sized estate agency chain. Lloyds were soon joined by other banks including Hambros and Trustee Savings Bank. The Building Societies Act (1986) and the Financial Services Act (1986) allowed building societies and insurance companies to diversify into estate agency, which they did with enthusiasm. The Halifax, Nationwide and Abbey National building societies set up their own estate agencies, as did insurance companies such as General Accident, Royal Life and Legal and General. Beaverstock et al. (1992: 171) state that:

> In each of these cases, the motives given for corporate expansion were essentially similar; to increase mortgage lending and to set up "one-stop property shops" which could offer a full range of property-related financial services (home contents insurance, life policies, pensions etc.). Banks, building societies and insurance companies justified their entry into estate agency business as the product of rational business strategies.

Unfortunately for most of these institutions, the timing and the methods of their expansion were badly misjudged. They chose to expand by purchasing existing estate agent chains, generally at a considerable premium, rather than establishing their own chains from scratch. As a result (Beaverstock et al., 1992: 173):

an acquisition war broke out in the estate agency sector and gathered pace after 1986. The large institutions were bent on buying as many estate agency firms as possible . . . takeover fever was heightened by the boom in the UK housing market . . . which meant that agency fee income rose to unprecedented levels. The buying spree undertaken by the large financial conglomerates became particularly frenetic in 1987 and 1988.

Beaverstock et al.'s (1992) figures suggest that the institutions collectively spent between £1 and £2 billion acquiring estate agency firms. They paid £300,000–£400,000 per branch, and a high multiple of annual fee earnings. Prudential Property Services, for example, paid £125 million for 337 branches, an average of £371,000 a branch. The cost of setting up a new branch from scratch was £75,000–£100,000.

Unfortunately for the institutions, the peak of the buying spree coincided with the peak of the housing market boom, which collapsed in 1989 in southern Britain. Fee income and profits slumped while overheads remained constant. As a result, most of the estate agency chains began to lose large sums of money and to retrench. In 1989 the 15 largest estate agency firms had losses of £170 million. Between December 1988 and December 1989 Nationwide closed 116 branches and Black Horse Agencies closed 41 branches. The most dramatic losses were suffered by the Prudential, which lost £23 million in the first half of 1990. It closed over 100 branches between December 1988 and May 1990, and another 175 in July 1990. In May 1991 it sold its property services division for £13.5 million, a 90% loss on the original capital invested. In some cases, individual estate agents were able to buy back their chains of offices for a fraction of what they were paid initially at the peak of the boom.

In January 1993 the Nationwide Building Society announced that it was closing a further 58 estate agency offices, bringing the total number of offices down from 510 in 1989 to 303: a reduction of 40%. This is in line with other cuts. General Accident had 590 offices at the peak, and 350 in 1992. Royal Life had 750 and was down to 500, and Hambro Countrywide had 520 offices in 1989 and went down to 450. The *Independent* (19 January 1993) stated that "Nationwide's estate agency network has never been profitable. It has lost a total of £50 million in its four years of operation." Perhaps the biggest indignity was the sale in 1994 of the remaining 300 Nationwide offices to Hambro Countrywide for £1 (Cicutti and Willcock, 1994). As Cicutti (1994) pointed out in an article titled "Estate agents are three a penny", the deal "brings to an end the costly and ill-judged attempt by building societies to diversify into selling houses, rather than just financing their purchase". And in 1995 the remaining 70 branches of Cornerstone, once Britain's largest privately owned estate agency, were put into receivership. Cornerstone was the result of a management buy-out from Abbey National, who had already lost over £240 million on the chain of 350 offices. The estate agency débâcle was a direct result of the financial liberalisation of the

1980s. It provides a good example of collective institutional misjudgement and the perils of expansion at the peak of a boom. The other example is the ill-judged foundation of a number of centralised independent mortgage lenders such as The Mortgage Corporation, owned by Salomon Brothers. Since 1994 five of the largest centralised independent lenders have been bought by the banks and building societies. Their key problem was that, by expanding at the peak of the market, they were forced to offer generous mortgage to income and advance to value ratios to attract business. But, not surprisingly, a higher proportion of their clients subsequently defaulted, leaving them with high levels of debt and no cushion of older safe mortgages to offset against the problematic new loans.

The Recovery of the Housing Market

The public perception of the housing market during the slump is very clearly shown through press cartoons. One of the most telling (Figure 2.17) was that in the *Independent* in September 1991, with an estate agent emerging from a bomb-damaged house (in a road of similarly damaged houses) saying: "I could have sworn I heard the all-clear". In the event, the all-clear did not sound until the end of 1993 in southern England when prices began to pick up again slowly. In the Midlands and the northern regions prices did not begin to pick up again until the beginning of 1996 and the national all-clear did not sound until mid 1995 according to the Halifax Building Society and early 1996 (Nationwide). Despite the gloomy predictions of some pundits such as Douglas Wood (1996), who saw prices remaining depressed for years to come, it is widely accepted that the continuing rise in real incomes and the sharp fall in house price/income ratios have made house purchase more affordable than at any time since the early 1980s (Marfleet and Pannell, 1996). The national house price/earnings ratio fell from a peak of 4.34 in 1990 to 3.39 in 1996.

As in previous booms, the recovery was led by London, particularly the central and inner London housing market, where a combination of rising salaries and bonuses in the City and high levels of international demand from South East Asian buyers began to force prices up sharply in 1996. The Nationwide Building Society indicates that average prices in London rose from £69,000 in the first quarter of 1995 to £95,481 in the third quarter of 1997: an overall increase of 38%. Average prices in London rose by 17% in the year to June 1997, 11% in the South East, 9% in the South West and 7% in East Anglia, falling steadily towards the peripheral regions, with annual rises of 3% in the North and 0% in Scotland. In some parts of inner London, prices have risen by over 25% in the last year, reminiscent of previous booms. As in previous booms, there is a "ripple effect" with prices rising in a wave outwards from London (Nationwide Building Society, 1997b).

Figure 2.17 "I could have sworn I heard the all-clear"

Source: The Independent, Friday 20 September 1991

At the time of writing, however, the overall consensus is that, with the exception of parts of London, where international demand is very strong, Britain is witnessing a slow recovery rather than a boom. Whether the recovery will turn into a boom remains an open question, although the continued rise in earnings over the past six years and the historically low house-price/income ratios which now prevail suggest that prices are likely to continue to recover. The ratio of house prices to disposable income, which averaged just 3.0 in London in 1993–96 compared to 5.5 at the peak of the boom in 1988, is comparable to the mid 1970s and the early 1980s (Nationwide Building Society, 1997a). The recovery is very uneven, however, in that the marginal first-time buyer properties such as one-bedroom studio flats, dubious conversions and cramped "starter homes" are being bypassed as a combination of lower property values, rising incomes and greater affordability now enables buyers to leapfrog the less desirable property and move directly into properties which could only have been bought by second-time buyers in the late 1980s (Earley, 1996). This is having the effect of leaving some marginal properties (and their owners) high and dry, unable to sell without taking quite considerable losses, if indeed they can sell at all. This is not a new phenomenon: marginal properties are drawn into the market at the peak of every boom, and are left high and dry as the tide recedes, but some marginal ex-council properties and converted flats appear virtually unsaleable or unmortgageable at present as lenders have tightened up their lending criteria (Forrest and Murie, 1995). There has been a "flight to quality".

In many respects, at the time of writing Britain is ripe for another house price boom as earnings have risen far above house prices since 1993, but several factors make the current situation different from that prevailing in previous recoveries. First, some economists such as Roger Bootle argue that we have entered a permanent low-inflation era and that previous assumptions from the 1970s and 1980s regarding house price inflation are now out of date. Because inflation is stable at around 3% per annum in Britain, as opposed to levels of up to 20% in the mid 1970s, the idea that house prices will rise as rapidly as they did in the past is now more problematic. Second, the 1980s boom means that many households are still carrying much higher levels of mortgage debt and this, combined with much greater uncertainty regarding house prices, is likely to lead to greater caution than hitherto. Certainly the evidence from the British Social Attitudes Survey supports this. The proportion of respondents supporting various statements concerning the desirability of house purchase and ownership has fallen significantly since 1989 and Curtice (1991: 103) stated:

> the popularity of owner occupation is conditional, not immutable. It appears to have been dented at least for the time being. . . . The attraction of ownership appears to wane as its potential reduces and its costs increase. . . . While it may provide autonomy and security, these attributes are apparently not enough to sustain its appeal when the economic advantages of ownership change.

What Triggered the Booms?

Explanations for the three booms vary but a number of main factors can be identified. These are (a) demographic shocks, (b) rising real incomes, (c) tax changes, (d) credit growth/liberalisation, and (e) expectations about future prices combined with the investment character of home ownership. The importance accorded to these factors varies considerably, but demography is a common element of many explanations. The immediate postwar baby boom hit the housing market in 1970 and the much bigger baby boom of the 1950s and 1960s coincided with the 1980s housing boom: the size of the 20–29 age group rose strongly throughout the 1980s (Cutler, 1995). Breedon and Joyce (1992: 174) state that "The upsurge in house prices in the late 1980s may have been partly generated by the growth in the population age group most likely to enter the housing market . . . increases in the size of this group have coincided with the two largest house price booms."

The second factor is the impact of sharp increases in real income. Real incomes rose rapidly from 1971 to 1973, from 1977 to 1979 and from 1987 to 1989. Real income growth was low throughout most of the second half of the 1960s but in 1972 and 1973 it grew 7–8% each year and average earnings rose much more rapidly. In 1978–79 real incomes rose by 6% a year and in 1988 and 1989 by 5% a year. Mortgage rates remained low at around 8.5% from 1970 to 1973. Stern (1992) argues on the basis of an econometric model of the determinants of UK house price inflation 1971 to 1989 that: "the principal driving forces behind nominal house price inflation are changes in real personal income and a one-period lag of house price inflation". He states that: "the housing boom of the early seventies was triggered by income rising faster in 1970 (3.9%) than in the previous two years, leading to a lagged response a year later, and by the dramatic fall in house building in 1969–70 (–27%) feeding through in 1972". Stern neglects, however, the role of demographic change in his analysis. He recognises the role of household formation but argues, without any evidence, that: "Old people trading down and growing family units should balance out on the whole in the UK's fairly stationary population" (Stern, 1992: 1328). He includes household formation as a variable but, as his analysis only starts in 1971, he is unable to detect the growth in new household formation in the early 1970s.

Lowe (1992: 73) suggests a fiscal basis for the first 1970s boom, arguing that:

> The abolition of schedule A taxation of owner occupiers in 1963 (the tax on the imputed rental income enjoyed by owner occupiers) in the context of the continuation of mortgage interest tax relief effectively created an incentive to home owners to use their house as an investment. Owners receive what is in effect a subsidy on the access payments but pay no tax on the imputed rental income nor on capital gains when they sell . . . from the mid 1960s, house prices began to rise above the RPI, due to . . . the capitalisation of these subsidies into house prices.

Lowe is correct about the abolition of schedule A taxation and the subsequent rise in house prices in the late 1960s, but arguably this marked the start of the long-term tax advantage of owner occupation over private renting (Hamnett and Randolph, 1987). The major boom in house prices came in the early 1970s, nearly ten years after the abolition of Schedule A taxation, which also had relatively little impact, as the ratable values on which it was set were long out of date. Nonetheless, tax changes can be important, and the Conservative abolition of the higher rates of income tax in 1987 contributed to rising real incomes in 1988 and 1989 and thus to house price rises in those years (Hamnett, 1997a).

Perhaps more important, however, for each of the three booms, were monetary policy and financial liberalisation. In the early 1970s the Barber boom created a wave of cheap credit, which made borrowing for house purchase easy, and Denis Healey did the same in 1978, bringing down mortgage interest rates to 9% after the 11% rates of 1974–77. In the early 1980s mortgage interest rates reached an all-time high of 15%, falling to a low in 1988 of 9%. Combined with the financial liberalisation of the early 1980s this had a dramatic effect. During the first half of the 1980s previous constraints on mortgage lending were greatly relaxed, the banks were able to compete more effectively in the mortgage market, the building societies abandoned their previous interest rate cartel, and they were able to borrow on the wholesale money market rather than being restricted to retail deposits. Breedon and Joyce (1992: 173) state that: "There was a marked increase in the average loan to value ratio in the early 1980s, as funds became more easily available. And second, and perhaps more importantly, unlike in previous house price booms, there was little mortgage rationing by building societies when house prices rose substantially in the late 1980s". This interpretation is strongly supported by Muellbauer (1990c) and Clapham (1996). But Cutler (1995) argues that while finance liberalisation is likely to have contributed to higher housing demand in the 1980s the main boom in house prices came several years after the deregulation of mortgage markets, which "suggests that other factors played a major part in generating the increase in housing demand between 1986 and 1989" (1995: 263). She points, in particular, to rising household expectations about future income and price rises and the impact this has on demand for assets such as houses. Milne and Clark (1990) suggest, however, that the impact of financial liberalisation was just deferred because of high interest rates in the early 1980s. Breedon and Joyce (1992) also stress the role of confidence and expectations, arguing that they "have been of fundamental importance in determining the demand for housing". They suggest (1982: 174) that:

By highlighting the role of expected capital gains, the asset market approach to house price behaviour helps to explain a number of features of the housing market. First, it is consistent with the observation that the volatility of house prices (up to 50% increases in one year) is more reminiscent of financial asset prices than of goods prices. Second, it allows for the anticipated effect of future events, such as changes in taxation.

In this context, Muellbauer (1990c) argues that there is evidence for a "frenzy" effect noted by Hendry (1984) in the context of the housing market, which operates when participants in the housing market become fearful of missing out on the rise or becoming locked out of owner occupation. But he adds that for a boom to begin, however, a shock is needed and "Financial liberalisation was the shock par excellence of the 1980s".

Breedon and Joyce (1992) produced an econometric model from 1970 to 1992 which suggests that "house prices are determined in the long run by incomes, wealth, user cost . . . the general level of prices, demography, financial liberalisation, and housing supply": in other words, all the factors discussed. It is certain that "the process of financial deregulation has set in train a long-term adjustment by the personal sector to new debt levels, and it would be reasonable to expect an upward adjustment in the long run house price to income ratio as a result" (Foley, 1992: 1). This is agreed by most observers (Miles, 1992a; Spencer and Scott, 1990), and Cutler notes that the average ratio of debt to income in the personal sector rose from 0.57 in 1980 to 1.17 in 1990. In other words, total personal debt including mortgages now exceeds income. Whereas home owners were forced to make large deposits when they purchased houses prior to 1980, and could only withdraw or borrow against housing equity with great difficulty, the liberalisation of mortgage lending has meant that households have rebalanced their debt/income/consumption ratios and have taken on, consciously or not, what may be a permanently higher level of mortgage debt. The implications of this shift, and the associated growth of equity extraction on consumer spending and the wider economy are discussed in Chapter 8. The future of house prices in the remainder of the 1990s and the early years of the next century are discussed in Chapter 9. Suffice to say here that Bootle (1995) and Cutler (1995) suggest that a more stable macroeconomic environment in the 1990s, with lower overall inflation, is likely to reduce the demand for housing as an investment and hedge against inflation. In addition, the continued demographic downturn in the key 20–29 age group points to a reduction of demand for home ownership. Finally, the erosion of confidence in the home ownership market caused by the slump may have increased the perceived risks. On the other hand, the house price/earnings ratio is now at a long-term low and rising incomes are likely to underpin another boom. The question is when, not if.

Appendix A: Problems of Measuring Price Changes

When trying to measure the impact of the slump on prices it is crucial to use one of the mix-adjusted house price series as these take into account changes in the mix of properties which are sold. If this were not done a change in the mix from, say, a large number of flats and a small number of larger, expensive houses, to a small number of flats and a large number of houses could have the effect of showing an overall rise in prices when prices fell for each type of property.

Discussing the mix-unadjusted DoE/BSA 5% sample survey of house prices, Fleming and Nellis (1983: 91) commented:

> Given the great heterogeneity of dwellings according to type, age, size and other physical and locational characteristics, to take this data source at its face value as a measure of price changes over time demands the strong assumption of constancy in the mix of transactions from one time period to the next. This is inherently unlikely and there is evidence to show that this is not the case – important shifts do take place in the transactions mix over relatively short periods of time.

The standard DoE/BSA 5% mix-unadjusted price series shows no price falls in any region until 1991, and then only very small ones. The figures for 1992 do show larger falls in the southern regions, but they are much smaller than either the DoE/BSA mix-adjusted sample, or the Halifax and Nationwide Building Society statistics, which indicate sharp falls from 1989 onwards. The problem is most clearly seen with the price data on Greater London. The DoE/BSA mix-unadjusted sample of mortgage completions shows average London prices rising from £77,697 in 1988 to £85,742 in 1991 and then falling to £78,254 in 1992. On the basis of the series prices rose steadily to the end of 1991. The DoE/BSA mix-adjusted series shows price falls from a peak of £98,700 in 1989 to £82,200 in 1993: a fall of 15.3%. But the Nationwide mix-adjusted data show that average prices in London fell by 25% between the first quarter of 1989 and the first quarter of 1992. The Halifax adjusted price series shows a similar fall. The discrepancies are similar for the other southern regions. The explanation suggested for this discrepancy between mix-adjusted and unadjusted price series is that there has been a sharp fall in the number of sales and prices of smaller, first-time buyer dwellings, but a relative increase in the proportion (but not the absolute number) of sales of more expensive property (albeit at much lower prices than in 1988). This change in sales mix would conceal the extent of price falls in both categories and overall in mix-unadjusted price data. It is thus suggested that mix-adjusted price series reflect what is happening more accurately than unadjusted price series when the market is in a severe downturn with sharp falls in sales volume and prices in particular segments of the market.

CHAPTER 3

Who Gets to Own? The Changing Social Basis of Home Ownership and its Implications

Introduction

Home ownership in Britain has undergone a remarkable expansion since the 1940s. In 1914 there were just 2 million owner occupied dwellings in England and Wales, and only 10% of households owned their own home. By 1938, after a decade of rapid speculative house building, the number of owner occupied dwellings had grown to almost 4 million and the proportion of home owners to 25%. By 1960, largely as a result of the sale of privately rented houses to tenants, the sector had grown to over 6 million and the proportion of home-owning households in Great Britain to 42%. The stock grew to over 8 million dwellings by 1970 and owner occupied households reached 50%. Today, there are some 16 million owner occupied dwellings in Britain (11 million owned on a mortgage, and 5 million owned outright) and 67% of households are home owners (Council for Mortgage Lenders, 1997; DoE, 1977).

The growth of home ownership has been part and parcel of a remarkable transformation of housing tenure structure in Britain. In 1914, private rented accommodation accounted for 80% of all households, and just 1% of households were public rented tenants. By 1945 ownership had grown to 25% and public renting to 12%, while private renting had declined to 54%. By 1980, public renting had climbed to its high-water mark of 31%. Subsequently, largely as a result of the "Right to Buy" policy introduced by the Conservatives in 1980, public renting declined to 19%, while private renting and housing associations accounted for 14%. In just fifty years the roles of the privately rented sector and home ownership have been completely reversed, and Britain has changed from being a nation of renters to being a nation of owners. As the dominant tenure, home ownership plays a crucial role in the British housing system. This transformation has been far less marked in other English-speaking countries such as the USA, Canada, Australia and New Zealand, where the dominance of home ownership was established far earlier as a consequence of the rural settler nature of these societies (Harris and Hamnett, 1987; Kemeny, 1981). It has also been less marked in Switzerland and Germany, where private renting is still very important, the cost of home ownership is greater and ownership is far less financially advantageous (Muellbauer, 1991).

Although it is often assumed that the growth of home ownership came about as a direct result of the desire to own, many other factors have played a role including rising income, the growing availability of mortgage finance, the decline of the private rented sector, the financial advantages of ownership, and the policies of Labour and Conservative governments which have reshaped the structure of housing opportunities (Hamnett and Randolph, 1989; Forrest, Murie and Williams, 1990; Saunders, 1990; Malpass and Murie, 1990). To attribute structural transformations of this magnitude to the outcome of consumer preference would be somewhat naive. If consumer preference for home ownership is so great today, why has it changed so rapidly and why are home ownership levels lower in some other, wealthier, countries? The answer is that the desire for ownership is not a simple, unmediated preference as some have assumed. Consumer choice takes place within the available structure of opportunities. When this changes, so do consumer preferences. This is not to deny the importance of choice and preference, but simply to argue that they are rarely unconstrained. As Henry Ford declared of buyers of the Model T: "They can have any colour they want so long as it's black." Government policy has been particularly important in encouraging owner occupation.

Government Policy Support for Home Ownership

Home ownership has long been seen, particularly by the Conservatives, as a desirable objective because it is thought to foster greater individual freedom and self-reliance, a greater degree of political stability, the potential for capital accumulation, and the ability to pass on accumulated wealth to children and beneficiaries. Labour, on the other hand, was primarily committed to the need for a greater level of social provision and only gradually came to accept home ownership as the majority tenure (Hamnett, 1987b, 1993b).

When Labour came to power in 1945, they faced a severe housing shortage – half a million houses had been destroyed by bombing and a further 3 million had been damaged. Not surprisingly, their priority was to direct the building industry towards repairs and to new social house building. Private building was rare during the 1940s. When the Conservatives came to power in 1951 (with Harold Macmillan as Minister of Housing) they continued a policy of council building and built more council houses in the 1950s (180,000 a year between 1951 and 1957) than any government before or since. But Macmillan came under strong political pressure and, when the worst shortages appeared to be over in the mid 1950s, the Conservatives switched to encourage new private building, focusing council building on slum clearance and redevelopment. The shift in Conservative policy towards owner occupation was clearly prefigured in the 1953 White Paper, "Houses: The Next Step" which stated: "One object of future housing policy will be to continue to promote, by all possible means, the building of new houses for owner occupation. Of all forms of saving this is one of the best. Of all forms of ownership this is one of the most satisfying to the individual and the most beneficial to the nation."

Slum clearance and redevelopment were to be the preserve of the local authorities, leaving general needs building to the private sector. This view was not restricted to the Conservatives. The evolving consensus about the limited future role of the public sector was reiterated in the 1965 Labour White Paper "The Housing Programme 1965 to 1970", which stated (para. 15) that:

Once the country has overcome its huge social problems of slumdom and obsolescence, and met the needs of the great cities for more houses let at moderate rents, the programme of subsidised council housing should decrease. The expansion of the public programme now proposed is to meet exceptional needs . . . The expansion of building for owner occupation, on the other hand, is normal: it reflects a long term social advance which should gradually pervade every region.

The Conservative White Paper "Fair Deal for Housing" (1971) stated that:

Home ownership is the most rewarding form of housing tenure. It satisfies a deep and natural desire on the part of the householder to have independent control of the home that shelters him and his family. It gives him the greatest possible security against the loss of his home, and particularly against the price changes that may threaten his ability to keep it. If the householder buys his house on a mortgage he builds up by steady saving a capital asset for himself and his dependents.

The next Conservative White Paper, "Widening the Choice: the Next Steps in Housing" (1973), reiterated that: "Most people want to own their own home. Their housing problems are best solved when they can exercise that choice. Many people who had no previous hope of home ownership now aspire towards it. The latent demand for owner occupation was stimulated by measures which the Government took."

In 1977 Labour's "Housing Policy: A Consultative Document" reiterated the view that home ownership was a "basic and natural desire": "A preference for home ownership is sometimes explained on the grounds that potential home owners believe it will bring them financial advantage. A far more likely reason for the secular trend towards home ownership is the sense of greater personal independence that it brings. For most people owning one's home is a basic and natural desire."

It is clear from these statements that both Labour and Conservative policy has been to foster home ownership as a natural tenure, with council housing being viewed as a necessary form of social provision for those households unable to achieve home ownership. Indeed, the Conservatives' desire to strengthen home ownership led to the policy of council house sales to sitting tenants. Margaret Thatcher (1979) referred to the Conservative council house sales policy as: "giving more of our people that freedom and mobility and that prospect of handing something on to their children and grandchildren . . . Thousands of people in council houses and new towns came out to support us for the first time because they wanted a chance to buy their own homes." She argued that the sale of council housing represented "a giant step towards making a reality of Anthony Eden's dream of a property owning democracy" (Hansard).

In 1990, the Conservative White Paper on housing stated that:

> Government's overall aim in its housing policy is that a decent home should be within reach of every family. Most people are now owner occupiers, and an important part of the Government's housing policy is therefore to continue to support the growth of home ownership through mortgage interest tax relief, the right to buy, and various low-cost home ownership schemes.

It has even been suggested (Malpass, 1990, Hamnett, 1987b) that Conservative housing policy was so ideologically geared to maximising home ownership and reducing council renting that it amounted to no more than a tenure policy, and analysis of Conservative policy changes during the 1980s and 1990s lends considerable support to this interpretation (Kemp, 1990; MacLennan and Gibb, 1990; Forrest and Murie, 1988; Hills, 1992). Subsidies for construction of social rented housing were dramatically cut back, while rents were increased, some 1.5 million social rented houses were sold to sitting tenants and the cost of mortgage interest tax relief rose rapidly during the 1980s. The Conservative Government was adamant that the council sector in Britain was too large and that municipal landlords were too dominant (Hamnett, 1993b). Consequently, in order to increase rental choice, council renting had to be cut back. These changes had considerable implications for the growth of home ownership. As Clapham (1996: 632) notes: "The spread of owner occupation to lower-income groups has been reinforced by the changes in the public rented sector. Local councils have been denied the opportunity to borrow . . . to build new housing. The size of the sector has declined considerably in the last few years."

In addition, the shift to a more market-orientated system resulted in the reduction of supply side subsidies, leading to higher rents. Council rents in England doubled from 6.9% of net income in 1980 to 12.7% in 1994. This made it difficult for those in work to afford the rents of new properties and social rented housing is becoming increasingly confined to those on social security benefits and Housing Benefit which covers rent costs. This, says Clapham, reinforced the social stigma of council housing and made it an unattractive option, financially and socially, for those who can enter owner occupation. Nor has private renting been a feasible or desirable option for many households. The sector comprises only 9% of the total and deregulation during the 1980s reduced security of tenure and abolished rent controls on new properties. Clapham concludes that: "The effect of government changes has been to reinforce the dominance of home-ownership in the British housing system . . . owner occupation is the only alternative for many households" (1996: 632).

Why does such a tenure transformation matter? The answer, at one level, is straightforward: the shift from private renting to home ownership has meant that the distribution of housing conditions, and opportunities, and financial costs and benefits has been profoundly altered. A nation of home owners is arguably a very different beast from a nation of private renters: economically, socially and politically. The financial implications are considered in Chapters 4 and 5 and the macroeconomic implications in Chapter 7. In this chapter the discussion focuses on the social and political dimensions of home ownership.

At the everyday level, it can be argued that home ownership is advantageous because it generally affords owners greater control over a crucial aspect of their own lives. Unlike tenants, home owners are commonly believed to have more freedom to choose (within financial limits) the sort of home they like, where they want to live, how they want to decorate their home, and so on. Tenants, on the other hand, are believed to have much less control and freedom of choice (Saunders, 1990), though Forrest and Murie (1990a) dispute many of these claims and suggest that tenants have considerable autonomy. In addition, the home is one of the most important aspects of people's lives. It is the base for domestic life and most people expend large amounts of time, money and effort to get the home they want, and to maintain and improve it. The home is a crucial social and cultural marker and, for some, an important source of social status and cultural distinction (Bourdieu, 1984). It is often a key element of personal identity (Saunders and Williams, 1988: Troy, 1991) and a major source of personal security (Cooper, 1976). Owners can renovate or improve their home more or less at will, and while this may be a source of frustration rather than liberation through DIY, it is important and owners may perceive a greater degree of personal freedom and autonomy than tenants. Irrespective of the academic arguments, survey evidence from the building societies and the Council of Mortgage Lenders (Pannell, 1997) shows that a consistently high proportion of the population identify home ownership as the preferred tenure, although in the 1990s the proportion has fallen, particularly among younger age groups, as a result of negative equity, repossessions, falling prices and lack of confidence in the market. The proportion choosing home ownership as their preferred tenure within two years has fallen from a peak of 81% in 1989 at the top of the housing market to 77% in 1996, and the proportion among the under-25 age group has fallen from 79% to 55%, although 85% of the age group expect to be home owners in ten years' time (down from a peak of 95% in 1989). But, as Pannell notes, it is unclear whether the reduction in interest of young adults is a permanent change or is "a temporary one related to the protracted recession in the housing market" (1997: 10).

The Changing Social Base of Home Ownership

One of the most important social consequences of the growth of home ownership in Britain since the Second World War has been its diffusion down the hierarchy of class and income. In the nineteenth and early twentieth centuries, the overwhelming majority of the population rented privately, and ownership was largely restricted to the middle classes (Burnett, 1986; Daunton, 1987; Kemp, 1982; Simpson and Lloyd, 1977), though working-class ownership was common in some areas of the country. It was not until the growth of ownership during the interwar years, especially in the South East, that home ownership became the dominant tenure for the middle classes (Jennings, 1971, Glass, 1963). It was the era when the suburban "semi" became common (Burnett, 1986).

The interwar evidence is fragmentary but a survey of middle-class households (local government officers, civil servants and teachers) in 1938/9 found that 18% of respondents owned their own homes, and 46% were buying them. Among local government officers the level of home ownership reached 70%. On the other hand, a Ministry of Labour survey of the families of insured workers in urban areas in 1937–38 (manual workers and non-manual workers with incomes below £250 per annum) found that only 18% were owners. Swennarton and Taylor (1985: 385) argue that in the interwar years: "only the elite of the working class could afford home ownership – and even then at the cost of self-sacrifice and thrift". Put simply, home ownership was out of range of most members of the working class given their low incomes relative to the price of houses and their uncertain employment. In a 1947 survey Gray found that 22% of British households owned their own homes and 4% were buying them. But, tellingly, the proportion of owners varied from just 15% among those earning less than £3 a week, to 39% of those earning between £5 and £10 per week and to 66% of those earning over £10 per week. There were variations from one town and city to another in the proportions of owners (Swennarton and Taylor, 1985) but the concern here is primarily with the national pattern.

The expansion of ownership to two-thirds of all households inevitably meant that it greatly widened its social base. Home ownership is now the mass tenure in Britain, as was private renting until the 1960s, and as a result it has become differentiated and fragmented (Forrest, Murie and Williams, 1990). What was once the preserve of the privileged few is now available to the many. Donnison and Ungerson (1982: 186) note that whereas: "Housing, income and class were once directly related to one another. The opening up of owner occupation to more than half the population – many in houses initially built for renting – has enormously diversified the range of people who buy their own homes, and the quality of the houses in this sector of the market."

The sources of growth have been various. While private builders have played a large part, constructing several million new houses, primarily in suburban or, more recently, in rural locations, a substantial share of the growth of home ownership is a result of transfers from private or public renting. Holmans (1987) shows that, of the growth of home ownership in England and Wales from 1939 to 1981 – a total of 8.2 million dwellings – no less than 3.4 million or 42% was accounted for by the sale of privately rented property for home ownership, compared to new building of 4.4 million or 53% and 5% from public housing.

And since 1981, over 1.5 million council houses have been sold for owner occupation. Forrest et al. (1994: 1) point out: "the substantial expansion of home ownership has been heavily dependent on tenure transfers from the council sector – accounting for 46% of ownership growth between 1981 and 1991. Such tenure transfers have been a major way in which home ownership has moved down the class and income scale."

This shift in social composition has had a number of implications. First and foremost, the expansion of home ownership to a level which parallels that in the USA, Canada, Australia and New Zealand is very important because it has opened up the benefits and the risks of home ownership to a much larger proportion of the population. Not only has home ownership given millions of people greater autonomy and control over their homes, free from the rules, regulations and dictates of landlords, but it has also opened up to large numbers of people the potential for financial accumulation. It cannot be too strongly emphasised that renters can pay for their dwelling several times over, whereas, all being well, owners can pay off the mortgage in twenty-five years or less and own their capital asset outright. Until recently, this was seen as an unqualified benefit, but the housing slump revealed the financial downside, particularly for marginal owners who had been drawn into home ownership during the 1980s by a combination of generous mortgages and low deposits and then found it difficult or impossible to meet mortgage repayments when interest rates rose, if they lost their job or experienced separation or divorce. As Forrest et al. (1996: 1) note:

> When prices began to fall in the late 1980s they impacted on a very different home ownership market than had previously existed. The downside of wider individual home ownership is that a greater mass of the population, and a greater proportion of those on lower and less secure incomes, are vulnerable to changing economic circumstances and the vicissitudes of interest rates.

Home Ownership and Social Class: The Changing Relationship

The changes in the social composition of housing tenures which took place as a result of the shift in tenure structures have very been considerable. Council housing was built under the provisions of the housing of the working classes acts until 1949 and initially, the high quality of new council housing compared to private renting attracted the skilled working classes who were able to afford the relatively high rents. This remained the situation until the 1950s when the growth of council housing meant that the semi-skilled and unskilled working class increasingly gained access to council housing. Meanwhile the skilled working class began to enter home ownership in large numbers, joining the growing ranks of the middle classes.

Table 3.1   Tenure by SEG of head of household

	Home ownership					Council renting			
	1961 %	1971 %	1981 %	1990 %	1961–90 ppc	1961 %	1971 %	1981 %	1991 %
Professionals and managers	67.3	75.8	82.7	90.3	+23.0	6.8	7.7	6.5	3.0
Intermediate and junior NM	53.4	59.3	70.5	78.4	+25.0	15.3	18.0	15.1	10.1
Skilled manual	40.0	47.8	58.4	73.0	+33.0	29.3	34.3	31.3	16.7
Semi-skilled	28.7	35.6	41.6	49.0	+20.3	32.3	39.2	41.9	24.2
Unskilled	21.9	27.0	30.9	38.0	+16.1	38.9	49.3	55.9	37.6

Source: 1961, 1971 and 1981 census of population; General Household Survey (1990)

Census data on housing tenure by socio-economic group (Table 3.1) show that since the 1960s home ownership has percolated down the social structure to a significant degree. In 1961 a majority (67%) of professionals and managerial households were home owners. The figure then fell to 53% of other non-manual workers, 40% of skilled manual workers, 29% of the semi-skilled and 22% of the unskilled. By comparison, just 7% of professional and managerial workers were in council housing in 1961 compared to 29% of the skilled and 32% of the semi-skilled manual workers, and 39% of the unskilled. Council housing was predominantly the preserve of the less skilled (Hamnett, 1984).

The proportion of owners in each group increased in subsequent decades and by 1991 no less than 90% of the professional and managerial group owned, as did 73% of skilled manual workers. By 1991, the top three groups had over 70% of households in owner occupation, and the semi-skilled approached 50%. The only group not to have achieved significant access to ownership by 1991 was the unskilled, although the proportion of unskilled workers in owner occupation almost doubled from 22% to 38% from 1961 to 1991.

Although marked differentials still remain, particularly between the top three groups and the bottom two, it is clear that home ownership has diffused down the social spectrum to a remarkable degree. Britain is now dominantly a nation of home owners and home ownership is no longer the prerogative of the middle class as it was in the 1950s and before. The proportion of the skilled manual group in home ownership has increased by a remarkable 33 percentage points from 40% to 73%, and this group is now numerically very important within the home ownership sector. Table 3.2 shows the composition of the main tenures by socio-economic group and economic activity status. Skilled manual workers comprise the second largest group amongst outright owners (12%), after the retired and economically inactive, and account for 30% of mortgaged owners, just below professionals and managers (34%).

Table 3.2 Socio-economic group and economic activity status of head of household by tenure, 1990

	Outright	Mortgaged	Council rented
Professional	2	10	0
Managers	9	24	2
Intermediate NM	4	14	2
Junior NM	3	7	4
Skilled manual	12	30	15
Semi-skilled	3	7	11
Unskilled	1	1	4
Economically inactive	66	7	61

Source: General Household Survey (1991)

The shift of the skilled working class and junior white collar workers into home ownership over the past thirty years has led to suggestions that housing tenures have become more sharply socially differentiated (Hamnett, 1984; Hamnett and Randolph, 1987; Bentham, 1986; Forrest and Murie, 1986) with the council tenure becoming increasingly socially residualised.

There are many different ways of measuring socio-tenurial polarisation or residualisation. It is possible to use socio-economic groups, income, age, the proportion of Supplementary Benefit recipients and other variables. Table 3.1 examined the changes in socio-economic group by tenure, and it is clear the council sector has attracted a growing proportion of the less skilled groups and has seen a reduction in proportion of the more highly skilled. Skilled workers have gradually moved out of council housing and into ownership while the council sector has become increasingly dominated by the unskilled, semi-skilled and, above all, the economically inactive. Forrest and Murie's (1988) comprehensive statistical analysis of residualisation and council housing showed that between 1980 and 1988 the proportion of households in each income decile who were council tenants changed dramatically, with an increase in the proportion in the lowest income decile and reductions in every other decile, with the falls increasing in size up the income scale. Conversely, the proportion of home owners in each income decile increased between 1980 and 1988 with the exception of the bottom decile, and increases were most marked in the top six deciles. The better off became more highly concentrated in home ownership and the less well off became increasingly concentrated in council housing. In 1980 council tenants comprised 17% of the bottom income decile, 15% of the second and 4% of the top decile. By 1988, the proportion of council tenants in the bottom two deciles had grown to 28% and 19% respectively, while the proportion in the top decile had fallen to just 1%. The proportion of council tenants in the top three deciles fell from 15% in 1980 to 5% in 1988. The council tenure is now dominantly the preserve of the poorest in society. Conversely, home ownership is now a more mixed tenure than it was thirty or forty years ago.

Table 3.3 Supplementary benefit recipients by tenure, 1967–87

			% of householder recipients in each tenure		
	Number (000)	Local authority tenants %	Tenants of private landlords %	Home owners	
				with mortgage %	without mortgage %
1967	2,174	45	37	3	13
1968	2,223				
1969	2,296				
1970	2,329				
1971	2,492	52	30	4	13
1972	2,475				
1973	2,292				
1974	2,285	57	25	4	13
1975	2,278	57	25	4	13
1976	2,346	58	24	5	12
1977	2,446	58	22	5	14
1978	2,420	60	21	4	14
1979	2,342	61	20	4	15
1980	2,462	61	19	5	13
1981	2,869	61	19	7	12
1982	3,208	62	18	7	12
1983	3,191	61	18	8	13
1984	3,389	61	17	8	13
1986	3,650	60	17	10	12
1987	3,684	61	18	9	11

Source: Forrest and Marie (1990) based on DHSS/DSS Social Security Statistics

These trends are reflected in the changing proportions of Supplementary Benefit recipients in the two main tenures. Forrest and Murie showed that in 1967 45% of recipients were council tenants, 16% were home owners and 37% were private tenants (Table 3.3). By 1987, the proportion of Supplementary Benefits recipients who were council tenants had risen 16 percentage points to 61%, the proportion of home owners rose 4 points to 20%, and the proportion of private renting fell 19 points to 18%. This shows that, as private renting has shrunk, the very poor have become concentrated in the council sector. Forrest and Murie also showed that while the proportion of economically active household heads declined from 70% to 60% between 1978 and 1987 across all tenures, it fell very sharply in the council sector from 60% to 38%, while the decline among mortgaged owner occupiers was from 97% to just 92%: a far smaller reduction. The council sector is increasingly the preserve of the poor, the less skilled and the economically inactive although home ownership has become more socially mixed and differentiated as it is now home to 67% of all households.

Who Owns What?

This finding raises an interesting question. If home ownership is now a mass tenure, comprising a variety of different social groups, to what extent is the home ownership market differentiated in other ways? One answer is that it is not home ownership *per se* which is important today, but the nature of what is owned. Although a growing proportion of the population have become home owners, the middle classes are likely to dominate the most desirable positions within the ownership market. The character of ownership as a positional good (Hirsch, 1978) may have changed significantly or become far more selective as home ownership has widened. Although there are far more working-class owners, they may be disproportionately concentrated in the less attractive or cheaper parts of the stock. Given the role of the price mechanism in allocating housing in the owner occupied sector this is what would be expected.

The home is also an important element of social display and a focus for the expression of cultural values and identity for most groups. It is frequently argued that this role is particularly important for the middle classes, not least because of its function as an indicator of social status and distinction (Bourdieu, 1984) especially in relation to those below. This was tellingly illustrated in George and Weedon Grossmiths *Diary of a Nobody* (1892), where Mr Pooter, a lowly City clerk, derives both satisfaction (and anxiety) from his six-roomed semi-detached house in the wastes of Holloway, North London. But others have generalised the analysis. In his analysis of gentrification in Melbourne, Jeager (1986) argues, after Veblen, that whereas the bourgeoisie occupy a strategic position, setting an example of conspicuous consumption, the middle class have to fight a war on two fronts between the dominant class and the "lower orders", from whom they struggle to demarcate themselves. He argues that an important way of doing this is via housing: "A change in social position is symbolised through a change in housing". Jeager argues that houses can express the same logic as conspicuous leisure or consumption. They designate the social rank and taste of their owner, and signify asethetic discernment and what Diggens (1978) terms "the cultural authority of wealth". Jeager argues that period terraced houses fulfil exactly this function for the middle-class gentrifiers: "Urban conservation is the production of social differentiation; it is a mechanism by which social differences are turned into social distinctions" (1986: 79).

Thrift and Leyshon (1991) have also argued, in a very different context, that the boom in the "country house" market in southern England in the mid to late 1980s stemmed from the rapid increases in income and wealth, particularly by those working in the City and financial services, and from the desire of those who had made money in the City or via Thatcherite entrepreneurialism to buy into the traditional "country gentry" lifestyle. And outside the country house market, people who are successful in business are likely to buy a house which symbolically mirrors their success through size and distinctive attributes. The

Table 3.4   Estimated current mean value of current property by SEG of head of household

	Mean value (£)	*n*
Professional	103,038	131
Managerial	103,519	600
Other non-manual	74,729	554
Skilled manual	66,351	717
Partly skilled	57,775	196
Unskilled	58,625	40
All groups	79,744	2,278

Source: British Household Panel Survey (1991)

carriage drive and porticoed entrance bear witness to the desire of "trade" to achieve respectability (Weiner, 1981).

The validity of this argument regarding class and house type can be shown using the first wave of British Household Panel Study (BHPS) data linking house type and house price and socio-economic group. This surveyed 5,000 households including 3,500 home owners in 1991. There is a clear relationship between current property value and the socio-economic group of the household. In general, professionals and managers own more expensive homes than other groups (Table 3.4). The value of their homes in 1991 was £103,000 compared to £74,700 for other non-manual groups, £66,000 for skilled manual groups and £58,000 for the partly skilled and unskilled. The differences hold by regions and over time and are unsurprising. Professional and managerial workers earn higher salaries than other groups and are likely to be able to pay more for a bigger, better or more expensive house. The relationship between socio-economic groups and house types using the BHPS data is very similar. A higher proportion of professional and managerial households live in detached houses (45% and 35% respectively) than any other group. Three times as many professionals as skilled workers (16%) live in detached houses, 4.5 times as many as semi-skilled workers (10%) and 6 times as many as unskilled workers (7%). Conversely, the representation of professionals and managers in terraced houses or flats (26%) is below other non-manual workers (50%), skilled workers (43%), semi-skilled (53%) and unskilled workers (59%). Semi-detached houses appear to be fairly evenly distributed (Table 3.5). While home ownership has grown, it is still differentiated in terms of price and type of house owned.

The results in Table 3.5 are broadly paralleled by the General Household Survey data on house type by socio-economic group of head of household, though this has the disadvantage that it is across all tenures. It shows that 43% of professionals lived in detached houses, compared to 38% of managers, 25% of intermediate non-manual workers, 17% of junior non-manual workers, 16% of skilled manual workers and just 4% of unskilled manual workers. The

Table 3.5 House-type by SEG of owner occupiers, 1991

	Prof.	Man.	ONM	Skilled	Semi	Unsk.	Total
Detached	45.3	34.9	16.0	15.6	10.3	7.4	21.3
Semi-det.	28.0	34.7	33.9	39.1	36.1	29.5	33.7
Terraced	16.1	18.6	29.5	32.3	38.7	45.2	32.6
Flat	10.6	11.8	20.6	13.0	14.9	17.9	12.4

Source: British Household Panel Survey (1991)

proportions living in flats show the reverse distribution. This conclusion reinforces Forrest et al.'s (1990) view that, as home ownership has grown, it has become differentiated and fragmented. Thus, distinctions within home ownership, particularly between marginal owners who own properties in poor condition or which are difficult to sell (such as "Right to Buy" owners and owners of flat conversions) and those living in desirable properties, may be as important as those between owners and non-owners. Karn et al. (1985) highlighted this in their study of inner city home ownership.

This conclusion should come as no surprise. In their book *Middle Class Housing in Britain*, Simpson and Lloyd (1977) highlight the importance of Birmingham's Edgbaston, the Park in Nottingham, Hampstead in London and Edinburgh's New Town as middle-class residential areas in the nineteenth century. These areas remain desirable and expensive. Indeed, for many members of today's expanded middle class they are unaffordable. A more realistic expectation for many of them is an inner city terrace or a four-bedroom "executive" house. The supply of desirable residences is always limited. As the middle class has grown in number, the type of property many of its members can afford is likely to have declined in quality, size and location from forty or fifty years ago, given the limited supply of large detached or terraced town houses in desirable areas. As the supply of desirable "positional goods" is relatively fixed, an increase in demand is likely to manifest itself in terms of a shift downmarket. Arguably this began to happen in the mid 1960s with the gentrification of some areas of London close to the centre, such as Primrose Hill, Chelsea and Barnsbury. It subsequently spread to areas such as Camden Town, Kentish Town and Fulham. By the 1990s the process had spread to Battersea, Stoke Newington, Whitechapel and Clerkenwell: all areas which would have been considered irredeemably downmarket in the 1970s. But, as the middle class grew in size and purchasing power, and existing inner city middle-class residential areas became increasingly expensive, expansion into hitherto marginal areas became almost inevitable. In my study of the flat conversion market in inner London (Hamnett, 1989) I found that a substantial proportion of new buyers were young middle-class couples or single people who were forced by rising prices to scale down their aspirations.

Home Ownership, Race and Gender

The expansion of home ownership has not simply been in income and class terms. Its gender and ethnic composition has also changed considerably over thirty years. Traditionally, home ownership was strongly dominated by male-headed households, although widows were prominent among outright owners. There were two principal reasons for this. First, single women were far less likely to attempt to buy their own home than men, not least because their incomes were generally much lower, and secondly, the building societies were generally very reluctant to lend to single women, or indeed to anyone other than a traditional male-headed household. The changes which have taken place were signalled in the Nationwide Building Society (1985b) report *Lending to Women* which showed that the proportion of mortgages going to women had risen from 8% in 1975 to 16% in 1985. One explanation for this change is demographic: the growing number of one-person households, but another is a change of attitudes. Building societies, which traditionally had been very wary of lending to single women (but not to single men) began to realise that women comprised a major market in their own right. Profits triumphed over prejudice. A recent study (Earley and Mulholland, 1995) points out that, though women comprised 51% of the population in 1991, only 24% of owner occupied households were female-headed and only 19% of female-headed households were buying on a mortgage, compared with 50% of male-headed households. Even among single people, the proportion of women with a mortgage (26%) is lower than men (35%), which reflects the sharp differences in occupations and earnings between men and women.

Certain ethnic minorities have also become much more strongly represented in home ownership in recent years, although this varies considerably between different groups. Whereas initially many New Commonwealth migrants were over-represented in the private rented sector owing to their lack of resources and their inability to gain access to council housing (Peach and Byron, 1995; Henderson and Karn, 1984), Indians, Pakistanis and African Asians have all gained a strong position in owner occupation in recent years according to the 1991 census and the latest PSI (1996) report. The 1981 census showed that, in terms of the country of birth of the head of household while 58% of those born in the UK were owner occupiers, this compared to 43% of those born in the Caribbean and to 74% of those born in Pakistan and Bangladesh and 77% of those born in India. The figures for 1991 show both country of birth and, more important, ethnicity (which takes account of the growing number of ethnic minorities born in Britain). In 1991, the proportion of home owners of white ethnic households was 68%, 43% of Caribbean ethnic origin, 77% of those of Indian, Pakistani or Bangladeshi origin, and 56% of those of Chinese ethnic origin. These figures confirm that those of Indian subcontinent ethnic origin have achieved a strong foothold in home ownership, whereas those of Caribbean origin tend to be more strongly represented in the council sector (37%) compared to those of both white (19%) and Indian (11%) origin. One of the main reasons for this discrepancy is the occupational class position of Caribbean households.

Home Ownership, Accumulation and Class Formation

This section considers one of the most important implications of the growth of home ownership: its potential for generating capital gains and capital losses. The significance of home ownership as a source of financial gains has been recognised over the years, but Peter Saunders raised a number of crucial issues regarding the links between home ownership and social class. Saunders argued that: "owner occupation provides access to a highly significant accumulative form of property ownership which generates specific economic interests which differ both from those of the owners of capital and from those of non-owners" (1978: 234).

This may seem fairly uncontroversial to non-sociologists, but its importance is that Saunders used this suggestion to argue that home ownership provides a basis for class formation separate from those based upon the ownership and control of the means of production or labour market position. Saunders's starting point was Weber, who argued that classes arose out of inequalities of economic power in commodity and labour markets. He differed from Marx in arguing that classes can arise in any market situation, not just in the relations between capital and labour. In Weberian terms, ownership of property constitutes an important basis for class formation, and Weber drew the distinction between property classes, whose members share common class situations by virtue of their command over forms of property that can realise income in the market, and acquisition classes, identified in terms of the marketable skills of different individuals. Saunders argued from this that home ownership functions, in Weberian terms, as the basis for a distinct property class separate from classes formed on the basis of labour market position. It is therefore possible to occupy one class position in relation to the production process and another in relation to the distribution of domestic property ownership. This, said Saunders, is not true of other consumption goods, which function for use rather than accumulation, and have no significance for class formation though they play a role in status group formation. Home ownership is unique as it functions both for use and accumulation, and provides the basis for a middle property class, distinct from classes formed on the basis of labour market position. As he put it: "if it can be demonstrated that house ownership provides access to real accumulation, then it may be . . . a basis for a distinct class formation in Weber's terms".

Saunders emphasised that this characteristic of domestic property ownership is not open to tenants who simply pay rent for the use of their property and cannot gain from any increase in its value. Home owners, by contrast, gradually pay off their mortgage and benefit from any increase in the value of their property. Saunders then argued empirically that house prices in Britain had increased faster than the general rate of inflation since the War, that over the ten years to 1978 they had risen faster than returns on other forms of investment, that mortgage interest rates had generally been negative and below the rate of inflation for many years, and that tax relief assisted the process of accumulation.

Saunders subsequently (1984) retracted his claim that domestic property ownership can be an element of class formation, on the grounds that class is based on ownership and control of the means of production and labour market position and that it is difficult to see where housing classes fit with classes based on labour market position. Instead, he argued that housing tenure and capital gains play a major role in social stratification in general, and that class is but one element of stratification. In many ways, Saunders's later argument was more radical than his original. He argued that class is not the only basis of social cleavage and stratification in contemporary capitalist societies, and that consumption sectors, of which housing tenure is the most important, represent an increasingly significant form of social cleavage which crosscut class divisions and may, in some circumstances, outweigh class divisions in importance. As he put it (1984: 207):

> Housing tenure, as one expression of the division between privatized and collectivized means of consumption, is analytically distinct from the question of class. It is neither the basis of class formations (as in the neo-Weberian tradition) nor the expression of them (as in the neo-Marxist tradition), but is rather the single most pertinent factor in determination of consumption sector cleavages. Because such cleavages are, in principle, no less important than class divisions in understanding contemporary social stratification. . . . it follows that . . . home ownership must remain central to the analysis of social divisions and political conflicts.

Saunders (1986: 158) went even further to argue that:

> Class is often a poor guide to a household's consumption location, for certain forms of private consumption are commonly purchased in many capitalist societies by large sections of the population, including many working class families . . . Consumption locations may generate effects which far outweigh those associated with class location. This must force us to reconsider our nineteenth century conceptions of class and inequality as phenomena of the organisation of production alone . . . Any analysis which insists on asserting the primacy of class is likely to achieve less and less understanding of patterns of power, privilege and inequality as these develop over the next ten or twenty years.

This is a radical claim which, if correct, greatly undermines the importance of class as a determinant of social inequality. Class, in Saunders's view, is just one dimension of social stratification, and it is of declining importance. Saunders's change of heart does not undercut his claim that home ownership provides a basis for accumulation and therefore functions as an axis for social stratification. His thesis has been the subject of considerable debate (Pratt, 1982; Ball, 1982; Harloe, 1984; Hamnett, 1989a) which will not be repeated here. Suffice to say that the main criticisms of his thesis are twofold. First, he failed to assess the degree to which the distribution of housing gains and losses are, in fact, related to labour market position and occupational class. Though Saunders accepted that position in the housing market is, to some extent, related to class and income, he argued that, in principle, gains from the housing market were independent of class. The debate hinges on the extent to which ownership (and the capital gains it offers) constitute a separate and distinct basis for class formation (or social stratification) from those deriving from the labour market and the ownership and control of the means of production. In other words, do gains or losses from the housing market reflect or reinforce differences in income and wealth derived from positions (past or present) in the labour market, or are they independent of labour market and social class? This issue is taken up in Chapters 4 and 5.

The second criticism is that the accumulative potential of home ownership is empirically contingent on a particular set of conditions which prevailed in the 1970s and mid 1980s. The force of this criticism was highlighted when the home ownership market in Britain entered a deep slump. Average house prices in Britain fell by 20% in nominal terms between 1989 and 1994 and in the South East of England they fell by over 30% and more in real terms. As a result, home ownership has been a source of considerable losses for many owners in the early 1990s rather than a source of accumulation. This is taken up in detail in Chapter 4. The argument may seem academic, and of limited importance. But it is not. If home ownership offers the potential of substantial gains and losses not related to position in the labour market, income and wealth, it comprises an important dimension of social stratification. We may need to add tenure to class, race and gender as a key dimension of social structure.

Home Owners and Political Alignment

Saunders's argument principally concerned the role of home ownership as a distinct basis for class formation independent of the labour market, but there is a related argument: because of the accumulative potential of home ownership, home owners can be seen as a distinct political alignment. Saunders argued (1978: 249) that "struggles between tenure groups reflect real divisions of political interest", and he suggested that a housing crisis "may well lead to a greater awareness of owner occupiers of the need to maintain their position in the face of possible threats". Pahl (1975: 298) also said: "It is when capital gains derived from housing advantage a whole class of society at the expense of another that it has socially divisive consequences. I see this tension between ownership and non-ownership increasing in the years ahead."

The argument about the potential political implication of home ownership has a long history and Conservatives have long argued that because owners own property they will be a force for social stability and against radical change. As Harold Bellman (1927) put it, home ownership was seen as "a bulwark against Bolshevism and all that Bolshevism stood for". Arguably, Margaret Thatcher's desire to give council tenants a right to buy owed as much to this sentiment as it did to any desire to open up the potential for capital gains to a wider cross-section of the population. Council housing was seen as a base for the Labour party, and the right to buy would weaken Labour support. The party argued that: "home ownership gives personal pride, and stimulates the natural instinct of care over, and preservation of, what is one's own. It helps create greater responsibility and stability in society" (Conservative Central Office, 1979).

If all this seems rather fanciful, it is appropriate to recall the events of January 1994 when the Westminster Council District Auditor published his report on the Conservative policy of selling off council houses for home ownership in parts of the borough to try to ensure that the social composition of the electorate in marginal wards was more favourable to the Conservatives. Shirley Porter, ex-leader of Westminster Council rejected the District Auditor's findings that five councillors be surcharged for the losses to the council incurred as a result of the sales policy. In December 1997 the High Court supported the District Auditor and found Porter guilty of lying. They upheld the surcharge of over £20 million pounds, but whether the money will be recovered is doubtful.

Not surprisingly, the far left has traditionally tended to be highly suspicious of home ownership, seeing it as individualistic and petit bourgeois, a way of reducing working-class solidarity, and aiding incorporation into a dominant ideology. Some analysts (Ball, 1983; Edel, 1981; Edel et al., 1984), have consequently argued that home owners do not make real gains or that working-class home owners do not benefit, and that ownership is not worth the candle. There is still an element of this kind of thinking on the far left, despite the fact that the great majority of skilled workers are now home owners in Britain and over 1.6 million former council tenants have opted to buy their own homes. It is, however, a mistake to assume, because home owners may have accumulated capital in the form of property, that they possess economic power in the sense that the owners of capital do (Forrest, 1983). As Friedrich Engels (1969) pointed out in "The Housing Question", a critique of Emil Sax who had argued that the solution to the housing problem lay in extending home ownership to the working class and transforming them into minor capitalists: "The worker who owns a little house . . . is no longer a proletarian, but it takes Herr Sax to call him a capitalist."

There is certainly historical evidence to support Conservative beliefs. Pattie et al. (1995) point out that in the General Elections from 1974 to 1992 about 50% of owners voted Conservative while only 20–25% voted Labour. By comparison, some 55–65% of council tenants voted Labour. In the 1983 General Election, for example, 44% of home owners voted Conservative and 16% voted Labour, while 18% of council tenants voted Conservative and 44% voted Labour: an almost mirror image between tenures. But the class composition of the two tenures is very different, and Pattie et al. note that the British Election Study found that since 1979 blue collar home owners are more likely to support the Conservatives than Labour. This ignores, however, the impact of the housing slump and the growth of negative equity in the early 1990s (an issue which is taken up in detail in Chapter 4), and Pattie et al. undertook a statistical analysis of the geographical shift in party support between the elections of 1987 and

1992 to see what effect the housing slump had on voting. To summarise a complex analysis, they found: "The higher the proportion of owner-occupied households locally with negative equity, the smaller the flows of support from other parties to the government, and the higher the flows from the government to the other parties ... there is strong evidence to suggest negative equity was a significant factor which lost the government support" (1995: 1311). They concluded: "After a decade in which government policy had advocated home-ownership as a means of achieving personal autonomy and financial security, it is hardly surprising that the evidence ... suggests that support for the government in 1992 flagged most where home owners had seen this dream turn sour" (1995: 1313).

This finding is very important because it suggests that the housing slump has undermined the idea that home owners have shared interests which may form the basis for a united political force. On the contrary, the differentiated nature of the housing market crisis, and the incidence of negative equity, meant that home owners reacted differently according to how they were affected. Those who lost out were more likely to vote against the Government, while those in areas where prices had risen, who were less affected by negative equity, were less likely to vote against the Government. Saunders may be right that tenure has political effects, but they are neither homogeneous nor easily predictable and may vary depending on when and where owners bought and their capital gains or losses, on the character of the housing market and many other issues. The problem of negative equity for the Conservatives in the 1997 election was foreseen by a number of commentators in 1995. The *Independent* (1995) ran an editorial entitled "Tory fears of repossession", which argued that many home owners in arrears had been sold a dream by the Tories which has turned into a nightmare. Will Hutton (1995) wrote in the *Guardian* that "Home ownership, embedded in folklore as a means of combining a roof over your head with a risk-free avenue to personal wealth, is instead becoming a liability ... just as an election approaches" and Tim Congdon (1995) argued in *The Times* that:

> From a Tory point of view, the plight of these households is a political disaster. Many of them saw home ownership as a passport to property owning democracy, the Thatcherite Britain of small-time capitalists. Even better, they thought, was the prospect that their admission to this happy land would be made easier by gratuitous capital gains from never-ending house-price increase. ... They might have expected that the Tory government, which had done so much to persuade them to maximise their mortgage debt and their financial commitment to market capitalism, would find a way to help them. Instead, it has actively penalised them. ... All they can do is grin ... struggle to keep up with the mortgage payments – and vote Labour.

Heath's cartoons in the *Independent* added their own comment (Figures 3.1 and 3.2).

Figure 3.1 Home ownership and party politics

Source: The Independent, Tuesday 6 June 1995

Figure 3.2 "New Civilisation Discovered"

Source: The Independent, Friday 20 December 1991

And this, it would seem, is exactly what many disgruntled home owners did, notwithstanding the attempt by Michael Heseltine immediately before the election to counter Labour party claims that the market was still suffering from negative equity, repossessions and stagnation (German, 1995). Indeed, Labour went on the offensive in March 1996, arguing the Tories were the "homewreckers' party" and had torn up the contract with home owners by cutting mortgage interest tax relief, raising taxes and adding to job insecurity (Rentoul, 1996) (Figure 3.1). Labour, said Tony Blair, was the party to "restore the shattered confidence of homeowners and those who want to buy". What is so remarkable about these claims and counterclaims is that the housing market slump handed Labour an unexpected weapon with which to attack the Tories – traditional party of home ownership – on its own ground. As Libby Purves (1995) noted in *The Times*, government ineptness was creating "a new social phenomenon: the property-owning underclass". To see the Conservatives undermined by the very thing they had sought to extend was to see them hoist by their own petard (Figure 3.2). Mass home ownership is clearly a two-edged sword, not merely a guarantee of Conservative voting as some believed. Home owners may be inclined to support the status quo if things are going well, but if the housing market is in a slump, those who believed that they were encouraged to buy with the prospect of almost guaranteed capital gains may be less generous.

CHAPTER 4
Winners and Losers:
Housing as an Investment

"A feature of the housing market in Britain is the enduring belief that home ownership is one of the best, if not the best, investment accessible to ordinary people" (Doling and Ford, 1991: 110). In the house price booms of the 1970s and 1980s the conventional wisdom that housing was a good, if not the best, investment, available to most people was strongly reinforced. Not only did nominal house prices keep well ahead of the rate of inflation, but building society finance was made available at relatively cheap and, until the 1980s, generally negative real rates of interest. Add on mortgage interest tax relief, the absence of capital gains tax on main residences and the absence of a tax on imputed rented income, and house purchase became what Kit Mahon, ex-Deputy Governor of the Bank of England, tellingly described as "a cheap, almost risk-free method of financing an appreciating asset with a depreciating debt". Whitehead (1979) commented that "Buying and living in one's own home has proved to be one of the most profitable investments, at least since 1945", and Ray Pahl (1975) observed in the wake of the early 1970s boom that "A family may gain more from the housing market in a few years than would be possible in savings from a lifetime of earnings." Nor were such views confined to academics. In a Parliamentary debate in 1979 on the proposal of the Thatcher Government to give council tenants a "Right to Buy" their own homes at a discount to the market price, Michael Heseltine, the Minister responsible, stated that:

In a way and on a scale that was quite unpredictable, ownership of property has brought financial gains of immense value to millions of our citizens. As house prices rose, the longer one had owned, the larger the gain became . . . this dramatic change in property values has opened up a divide in the nation between those who own their own homes and those who do not.

This view was supported by the Nationwide Building Society (1986), which pointed out in *Housing as an Investment* that buying a house to live in had proved a better investment than any other over the long term on a variety of measures. Taking a range of dates from 1960 onwards when a house was bought, it first looked at average annual changes in house prices to 1985. Second, it looked at the average gross rate of return on a house bought with a twenty-five-year mortgage and sold in 1985, taking into account transaction costs (such as estate agents' fees, legal changes and stamp duty) and gross (before mortgage interest tax relief) mortgage repayments at the prevailing basic mortgage rate. Then it looked at average net rate of return after taking into account mortgage interest tax relief at basic rate, and finally it included the notional benefit derived from living in a house rent-free.

The report concluded first that, except for the short term, buying a house on a mortgage gave "gross and net rates of return which are into double figures: consistently exceeding the rate of inflation". Secondly, it suggested that if the notional value of living in a house rent-free was taken into account, the net rate of return is greatly increased, rising to 25% per annum for a house bought in 1970. Thirdly, it argued that even gross rates of return on home ownership were higher than those of other financial investments, and that if mortgage tax relief is taken into account (which creates an advantage for owners but lowers the rate of return for most other forms of saving on which the interest is taxed), the difference was greater still, comfortably exceeding the returns on various savings accounts and exceeding the returns on unit trusts and the FT ordinary share index. The report also noted that, looking at the ten years 1975–85, the annual rate of return varied over the country, and that on an 80% mortgage advance net returns ranged from a high of 17–18% in London and the Outer Metropolitan area to 15–16% in East Anglia, the Outer South East and the South West to between 11–13% in most of the rest of Britain. Surprisingly, the report concluded that the annual rate of return to owners who move to a more expensive house every five years (25% above average price on the first move, 50% on the next) are lower than those who did not move house, taking transactions costs into account. This contradicts the common assumption that moving upmarket is a way of maximising capital gains, and is discussed later in the chapter.

Nor were such views specific to Britain. Anthony Downs (1980) suggested that in the USA an increasing number of households had realised that house price inflation had increased the financial advantages of moving their savings "under their own roof". He argued that as people could earn greater rates of return on real estate investments – if they borrow most of the money – than they can on savings accounts, they "have come to look on borrowing to purchase housing as the best way to increase their family savings. They believe, as my father used to say, that 'what you owe today, you will be worth tomorrow'." One of the strongest academic proponents of home ownership as a source of capital gains is Saunders (1990), whose views were outlined in Chapter 3. Although he conceded that his "analysis was founded on certain empirical conditions, regarding the potential for accumulation", he stated that:

> The returns which owner-occupiers have been making on their housing compare favourably with . . . other investments and far outstrip the general rate of inflation . . . this is likely to continue in the long term . . . real gains have been made through most of this century and certainly ever since the 1950s . . . there is little prospect in Britain of the housing market collapsing as it did in the Netherlands (1990: 202–3).

Real house prices have risen over a long period in the past, and the likelihood is that they will continue to rise in the future. There is no reason to accept the arguments of those who claim that the gains will prove temporary and that owner occupiers are about to get their comeuppance (1990: 154).

As we now know well, the 1990s slump completely undermined Saunders's case. The housing market in Britain did collapse as it had in the Netherlands and owner occupiers in London and the South East got their comeuppance. But even before the slump Saunders's analysis had been challenged on a wide variety of theoretical and empirical grounds by various authors (Hamnett, 1989a; Harloe, 1984; Forrest et al., 1990; Murie, 1991). One of the most important criticisms, from the point of view of this book, is that the debate on capital accumulation has been unduly based on the experience of the home ownership market in Britain during the 1970s and 1980s, when house price inflation was particularly marked. Thorns (1989) argued that when Saunders's arguments are viewed in the light of the Australian and New Zealand experience "A number of significant differences arise which raise doubts about the transferability of the argument beyond the confines of the British experience" (1989: 214). He argued that Saunders's thesis should be set against a more extensive time period than the last few decades, adding that:

When the experience of home ownership over such a longer period of time is considered, questions are raised about the inevitability of the process of capital accumulation within housing. Data from both the USA . . . and New Zealand show that the history of accumulation has been one of fluctuations in both the rate of gain and its extent over time. There have in fact been times when housing would have produced losses as well as gains. How far this is the case in Britain is still open to debate (1989: 216).

The 1990s slump in the British home ownership market closed the debate. It is now clear that home ownership in Britain can produce widespread losses as well as gains. But others had argued a more sceptical position during the 1980s and Peter Spencer (1987) claimed that although house price inflation was brisk during the 1980s, when improvement expenditure, changes in real interest rates, maintainance costs, rates and imputed rent were taken into account, the net return of owner occupation was negative during most of the 1980s until 1986. He concluded that although owner occupation was an attractive financial proposition during the 1960s and 1970s, this was not true of the first half of the 1980s. This view was backed by Duncan (1990), who claimed that house prices do not rise consistently or uniformly. He argued that, when prices are measured on the basis of comparable quality and quantity, many of the price rises are ficticious and that "Falls are almost as common as rises. In the medium-term increases are significantly less in Britain than are supposed and in the long-term may be ephemeral for home-owners as a group" (1990: 31).

This conclusion contradicts most of the long-term evidence from Holmans (1990) and others on long-term house price changes, which show increases of 200% in real terms from 1966 to 1989. Duncan's view can be said to be over-stated, but his scepticism regarding house price inflation proved prescient in the slump which followed. As Robin Leigh Pemberton (1993), Governor of the Bank of England, put it in a speech to the British Property Federation:

> Home ownership was seen as a laudable social goal, especially as post-war purchasers saw the value of the investment in their property rise . . . The effect of growing demand was to drive prices upwards more rapidly than inflation: this added the final twist. Property became seen not only as the best hedge against inflation, but even as a form of savings which offered high, sometimes very high, returns which dwarfed and ousted other savings opportunities. This forced prices up even more quickly, which in turn stimulated demand as purchasers extended themselves financially in order to participate as much as possible in a game in which it then seemed no one could lose. We know only too well now what trouble this game could bring, especially to purchasers who now find themselves owing more than their property is worth at a time when many families face hardship through reduced earnings and unemployment. In the long run it must be no bad thing that prices should be more realistic – it will be much to the advantage of new first time buyers. But that is no consolation to those caught when the market peaked.

This points to the role of time as a key determinant of capital gains and losses: an argument Thorns (1989) has made in the context of New Zealand. He points out that in New Zealand variations in capital gain depend "on the point in the cycle of booms and slumps that the household enters and exits from the housing market" (1989: 296). This point is reinforced by Dupuis (1992), and common sense suggests that buyers who purchase before the peak of a boom are likely to do better than those who buy at or near the peak. So too, the longer one has owned, the greater the gains are likely to be.

The Calculation of Gains and Losses

It is possible to calculate the gains (and losses) from home ownership in a variety of ways, and there is much dispute over the most appropriate way to measure gains (Dupuis, 1992; Thorns, 1989; Saunders, 1990; Badcock, 1989). Some authors (Saunders, 1990) stress the importance of relative gains, and rate of gain which takes into account time owned, while others (Murie, 1991) argue it is better to focus on absolute gains, as a large percentage gain measured on a small base is meaningless. Saunders (1990) also argues that the most appropriate way to calculate gains is on the basis of the deposit, as this constitutes the

Table 4.1 The range of potential gain measures

First property				Current property			
Nominal		Real		Nominal		Real	
Absolute	Relative %	Absolute	Relative %	Absolute	Relative %	Absolute	Relative %

Illusionary Gain (current estimated value – price paid)
Crude Gain (current value – price paid – any outstanding mortgage)
Gross Gain (current value – price paid – any outstanding mortgage – deposit)
Net Gain (current value – price paid – any outstanding mortgage – deposit – cost of
  improvements)

owners' original capital investment. Murie, however, argues that, while logical, this approach: "neglects the fact that the rate of return ... does not derive only from the deposit. It only occurs because of the accompanying loan" (1991: 356). Murie is also extremely critical of the use of the deposit as a base for calculation as it "produces an idiosyncratic pattern of rates of return" which rises as the size of the deposit falls. In the case where 100% of the purchase price is borrowed, the rate of return is infinite. Thus, this method of calculation leads to very high rates of gain (or loss) where the deposit is small.

A variety of different calculations (nominal and real) based on purchase price can be used to capture the range of experiences. It is possible to think of these in terms of a dual hierarchy of growing complexity and sophistication. There is a dual hierarchy because it depends on whether the calculations are based just on the current home or whether they go back to the first home owned, which gives a measure of gains and losses over the whole ownership career. There is also a difference depending on whether the focus is on the absolute gain or loss or on percentage gain or loss, based on initial purchase price. Finally, calculations can be based on nominal gains or on real gains, which take inflation into account (Table 4.1).

The first, the simplest, and in some ways the most misleading, form of calculating gains is to take the difference between the current value of the current home and the original cost of the current home. This can also be calculated for the original cost of the first property owned. But, because this calculation does not take into account outstanding mortgages, deposits, the cost of improvements or equity extracted, we have termed this the illusionary gain. It is akin to what Dupuis (1992) terms the capital gain. As she points out:

Although capital gain is particularly useful when making comparisons between house price inflation both within and between areas, its major limitation is that all the owner's financial input, such as the mortgage debt, mortgage payments, renovations, repairs, maintainance, the costs of buying and selling and any other costs that are incurred in the normal course of owning a house, is ignored. It describes, therefore, the unlikely situation of owners living in a house mortgage and cost free (1992: 29).

The second measure of gain, which is termed crude gain, deducts any outstanding mortgage on the home. The third measure of gain, gross gain, also takes into account the deposit on the property. Dupuis terms this the wealth increase. It is possible to think of more sophisticated calculations which take into account the cost of major renovations and repairs (which we term the net gain), and equity extracted from the home. Finally, it is possible to calculate home owners' equity, by subtracting any outstanding mortgage debt from the current estimated value. This is not, however a measure of gains, strictly speaking, but of housing wealth. This is discussed in Chapter 5.

Because of the way in which the different calculations of gain and loss were constructed, the illusionary measure of gains always shows the greatest gains and the smallest losses, as it ignores both deposit and outstanding mortgage. Crude and gross gains are smaller (and losses greater) as they also take into account mortgages outstanding (crude) as well as the deposit (gross). Real gains are always smaller than equivalent nominal gains because of inflation.

The falls in nominal house prices in Britain noted in Chapter 2 are indicative, but they do not show the actual distribution of gains and losses amongst individual households. For this, it is necessary to draw on direct evidence. As there is no publicly accessible data source on individual house price changes over time in Britain, I have drawn on two sources of survey data: the first wave of the nationwide British Household Panel Survey of 3,672 owner occupied households carried out in 1991, which Jenny Seavers and I analysed as part of a Rowntree Foundation-funded project on Income and Wealth Inequality in Britain; and a MORI survey of 972 home owners in five areas of the South East of England in late 1993 and early 1994 which formed part of an ESRC-funded research project on home ownership carried out at the Open University by Phillip Sarre, Jenny Seavers and myself.

Distribution of National Gains and Losses:
The British Household Panel Survey

To provide a wide picture of the distribution of gains and losses both in the South East and in Britain as a whole we analysed data from the first wave of the British Household Panel Survey. This survey employs a stratified random sample of 5,500 households in Great Britain (excluding Northern Ireland). Of these, 3,672 (67%) were home owners and we had a total of 3,165 valid cases for 1991. As with the MORI survey, BHPS respondents were asked to estimate the current market value of their home and to give price paid; this forms the basis of calculation. There were some difficulties concerning the calculation of outstanding mortgages (see Hamnett and Seavers, 1996 for details), and endowment mortgages were treated as repayment mortgages in calculating the proportion of

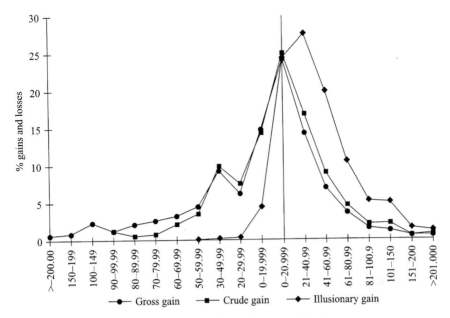

Figure 4.1 Percentage distribution of gains and losses, 1991 (£'s)

Source: British Household Panel Survey (1991)

the mortgage which has been paid off. Given the low value of endowments if they are cashed in prematurely, this will greatly understate the scale of losses and overstate gains, particularly where owners have only had endowments for a few years. The principal difference between MORI and the BHPS is that BHPS asked only about the current home. It is thus possible to calculate BHPS gains and losses only for the current home, and not for the first home.

Looking first at the distribution of gains and losses for the UK as a whole, the BHPS shows (Figure 4.1) that the distribution of illusory gains and losses was positively skewed with a median value of £21,000–41,000, with 13% making gains of over £81,000: 5% of households had made losses, the majority of which (4.4%), were under £20,000. Crude gains were normally distributed with a median gain of £0–21,000. The proportion of losers was much greater than for illusionary gains: 40% of home owners had made losses: 22% had losses of under £30,000 and 18% had losses of £30,000 or over, including 8% with losses of over £50,000. The distribution of gross gains was negatively skewed with a median value of £0–21,000 and 47% of owners had made losses, including 17% who lost over £50,000.

The national figures show that Saunders's thesis regarding the inevitability of gains from home ownership was empirically incorrect in the early 1990s. But, as we have shown earlier, house price falls in the 1990s were most strongly marked in the South. This suggests that the proportion of losers will be much higher in

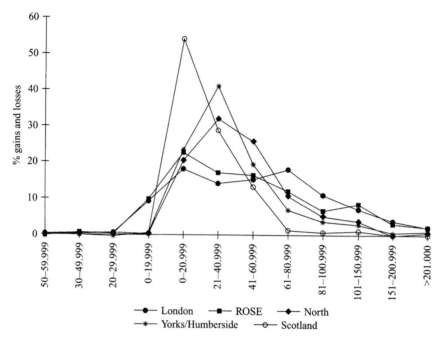

Figure 4.2 Percentage distribution of nominal illusionary gains and losses by region, 1991 (£'s)

Source: British Household Panel Survey (1991)

the South and regional analysis of BHPS data supports this. Taking illusionary gains, the proportion of losers varied from 11% in London, the rest of the South East and East Anglia to under 1% in the northern region (Figure 4.2). But, as Figure 4.2 shows, the distribution of gains in the North and Scotland is much more tightly peaked than in the southern regions. In Scotland, 54% of households had gains of less than £29,999, 83% gained less than £40,999 and 96% less than £60,999. By comparison, the distribution of gains in the southern regions is much more widely distributed. Only 32% of London households and 40% in the rest of the South East (ROSE) were in the £0–40,999 category whereas 24% of households in London and 20% in the South East gained over £81,000. Though the proportion of losers is higher in South East than elsewhere, the proportion of high gainers is much greater than in the northern regions. This reflects higher house prices in the South, and indicates that if prices had not fallen sharply in the South in the early 1990s the proportion of high gains would be even higher.

The distribution of crude and gross gains is similar, as the southern regions have a much higher proportion of losers than the northern region but higher proportions of high gainers. For crude gains, the proportion of losers was 49% in London and 47% in the South East compared to 35% in the North. More striking, however, is the proportion of those losing over £20,000. This ranged from 42% in London and 36% in the South East to 11% in the North. In the house price roller coaster of the late 1980s and 1990s, a higher proportion of home owners in the South lost out, but those who bought earlier or avoided paying high prices at the peak of the boom still gained more than elsewhere.

The Distribution of Gains and Losses in South East England

The five areas we chose for our surveys included two London boroughs (Hammersmith and Haringey) and three other areas (Milton Keynes, Oxford and Chiltern Urban District Council). The areas were selected on the basis of a number of criteria. We wanted an inner and an outer London borough, an authority in the Outer Metropolitan Area (OMA) and the Outer South East. Secondly, we also wanted areas which were representative in terms of the distribution of house prices and house price inflation in the region during the 1980s. Households were asked a range of questions concerning their current and previous homes, including purchase price, current estimated value, deposit and mortgage outstanding to provide information to calculate gains and losses. The reliance on estimated house prices by respondents is a weakness in the methodology but it was impossible to get around it. Insofar as our respondents may have underestimated the extent of the fall in prices since 1989, our results will underestimate the incidence and scale of losses. The same is true of the BHPS, which asked respondents to estimate the current value of their house.

Because our survey of home owners was undertaken in late 1992 and early 1993, at the bottom of the housing market slump, it is not surprising to find that home ownership is not just a way to accrue capital gains, as Saunders assumed in the 1980s. It has also functioned as a source of capital losses, particularly for those people who bought in the mid to late 1980s near the top of the market only to see the price of property fall back sharply, particularly in the South of England. We found that, even using the estimated current market value (which may be an overestimate), 8% of respondents had experienced a nominal illusionary loss based on the purchase price of their first property, and 20% had experienced illusionary losses on the current home. The difference is not surprising, as those who have owned one or more previous homes are likely to have been in the home ownership market longer than those in their first home and will have generally bought for far less than the current value (Figure 4.3).

Looking at nominal crude gains and losses (i.e. taking outstanding mortgage into account), the distribution of winners and losers is very different to that of illusionary gain. Some 44% of owners had experienced losses based on the price of the current home and 28% on the price of their first home (Figure 4.4). The proportion of losers calculated on the basis of nominal gross gains is greater still: 46% of households incurred losses based on their current home and 31% on their first home. As this is arguably the most realistic measure of gains and

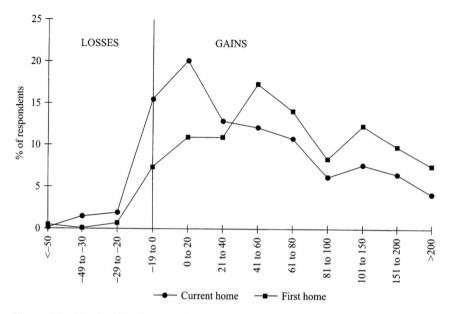

Figure 4.3 Nominal illusionary gains and losses on first and current homes (£'s)
Source: South East Homeowner's Survey (1993)

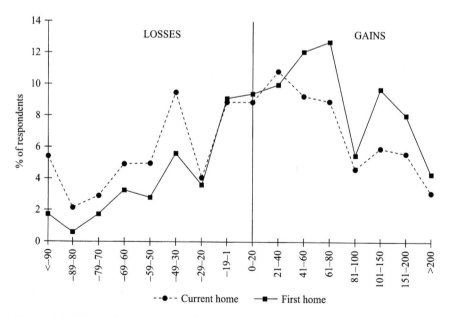

Figure 4.4 Nominal crude gains and losses on first and current homes (£'s)
Source: South East Homeowner's Survey (1993)

Table 4.2 The proportion (%) of households experiencing losses on the basis of their current or first property

	Current property	First property
Nominal Illusionary Gain	19.8	8.1
Nominal Crude Gain	43.7	27.7
Nominal Gross Gain	46.1	30.6
Real Illusionary Gain	37.4	17.5
Real Crude Gain	57.6	40.4

Source: South East Homeowner's Survey (1993)

losses, including both outstanding mortgage and deposit, the high proportion of losers totally undermines Saunders's thesis regarding the inevitability of gains from home ownership.

Illusionary losses were relatively small when measured on the basis of the current home (16% experienced losses of between £1,000 and £16,000 compared to 0.2% who lost over £50,000), losses were much greater using crude and gross measures. On the crude gains measure 14% of owners had losses of £20,000–49,000, with 13% losing £50,000–79,000 and 7% losing £80,000–99,000. On the basis of gross figures, 11% lost over £80,000 and 6% had losses of over £100,000.

When inflation is taken into account and real gains and losses are analysed, the proportion of losers is even greater. Where real illusionary gains are concerned, 37% of owners experienced losses based on current property (Figure 4.3), and 17% had losses based on first property. When outstanding mortgage is included (real crude gains) 58% of households experienced losses on current property and 40% had losses on the basis of first property (Table 4.2 and Figure 4.3).

These figures are remarkable. They show that, rather than home ownership being a consistent source of capital gains as Saunders (1990) suggests, the slump in prices in South East since 1989 has meant that, taking outstanding mortgage and deposit into account, nearly half the owners in our sample lost money on their current home and 30% lost even on the basis of the price of their first home. Measured in real terms, the proportions are even greater.

Three criticisms could be made of these findings. First, it may be objected that the figures reflect a period when house prices fell substantially, and are thus unrepresentative. Such a criticism would be misguided on two grounds. First, Saunders's theory regarding capital gains was founded on the view that house prices in Britain had risen since the 1950s and were unlikely to fall. This has been shown not to be the case. Second, the figures are based on a stratified random sample of owners, including those who bought their houses many years ago. They represent a cross-section of all owners in the region. As house prices in the South East recover, which they began to do in 1996, the proportion of losers will clearly fall, but in late 1992 and early 1993, when the survey was carried out, a large proportion of owners had experienced losses. A third criticism could be that our figures are unrepresentative as they are based on the South East. This criticism is groundless as it was precisely in the South East that prices previously rose most rapidly. The existence of widespread losses anywhere at any time period is a crucial counter to Saunders's thesis.

Ownership in Britain was a major source of capital gains during the 1960s, 1970s and most of the 1980s. But this broke down in the 1990s as the home ownership market went into a prolonged slump. It would appear that Saunders's thesis held true only for a historically specific set of conditions. Whether this shift will prove to be temporary or permanent remains to be seen. Many commentators are of the view that in a new low inflation regime combined with a severe demographic downturn in numbers of households in their twenties, the conditions which led to house price inflation in the 1970s and 1980s have now ended. The house price slump has also administered a shock to the generation of new home owners who entered the market in the late 1980s or after. Their experience of ownership is one of falling prices and their expectations will be very different to those of previous cohorts of home buyers.

The Determinants of Gains and Losses in the South East

As we showed in the previous section, the distribution of financial gains and losses from home ownership in Britain is highly uneven. Depending on the measure chosen, the proportion of losers ranges from 8% to almost 60% in the South East and from 5% to 47% in Britain overall. Equally, the size of gains and losses ranges from hundreds of pounds to hundreds of thousands of pounds. The crucial issue examined in this chapter is how these gains and losses are distributed among different groups of home owners: who gains and who loses? What are the main correlates or influences on the distribution and scale of gains and losses? Are some categories of home owners more likely to gain than others and, if so, why? As Thorns (1989: 294) puts it: "the wealth accumulated via ownership in the form of capital appreciation is essentially money that has been redistributed through the operations of the property market. The key research questions therefore became what are the mechanisms of this redistribution and who were the winners and the losers."

The traditional view is that the strongest influence on the scale of absolute capital gains is social class and income, primarily because the higher social classes and income groups are more likely to purchase more expensive property, to have bought earlier and owned for longer, to have moved more often and to have moved up the housing ladder to more expensive houses, which tend to give rise to larger absolute capital gains than cheaper houses. If, as is sometimes argued, more expensive houses increase in price at a faster rate than cheaper property, the absolute gain will be disproportionately larger. But there is considerable dispute over the importance of social class as a determinant and predictor of the scale or rate of capital gains from home ownership. On the one hand, commentators such as Murie (1986, 1991), Forrest and Murie (1989a) and Forrest et al. (1990) point to the importance of class and position in the paid labour force as a major influence on the scale of gains. Saunders (1990), on the other hand, argues that social class is not a predictor of capital gains, gains from home ownership do not favour the middle classes and, in relative terms, the rate of gains is higher for lower classes. This has been strongly challenged by Murie (1991) and Forrest et al. (1990), who argue that, given the distribution of home ownership by type and value by class, absolute gains are likely to be strongly class-related: higher classes making larger absolute gains. They have also criticised Saunders's (1990) claims that, measured in relative terms, working-class owners generally have higher proportionate gains than middle-class owners. This is said to result from Saunders's method of calculating gains on the basis of the deposit. Because many working-class owners, particularly "Right to Buy" owners, put down low or zero deposits, they make far larger proportionate gains.

But class and income are not the only factors which can influence the scale or rate of capital gains. As argued above, gains are generally related to the price of property: the more expensive the property, the greater the absolute gains. Second, the distribution of gains will be related to the geographical differences in house prices and rates of price increase or decline. There is strong evidence from Britain and other countries that rates of price increase vary sharply from area to area and region to region at different times (Harris 1986, Hamnett, 1982, 1993a; Dupuis, 1992; Thorns, 1989). London and the South East have consistently led house price booms in Britain and, early in a boom, the regional house price gap between North and South is generally at its greatest, falling in subsequent years as prices in the North catch up.

Finally, it is agreed that, where prices rise over time, the longer owners have owned the greater the likely gains. But, as we pointed out in the previous chapter, given the slump in the British home ownership market since 1989, it is now appreciated that buying when prices are at a peak or falling can lead to capital losses. In sum, a number of factors are likely to influence absolute and relative gains and losses from home ownership. Thorns (1992: 216) argues that "patterns of accumulation are variable rather than uniform", and argues that there is a "need to move from an aggregate level of analysis to one which takes greater recognition of when a house was bought, its location, how long it was owned and the social characteristics of the owner". Similarly, Dupuis (1992: 43) argues from New Zealand evidence that "the extent of financial gain is strongly dependent on both geographical location and the time period for which the gain was calculated". This points to the role of time as a key determinant of capital gains and losses; and Thorns (1989: 299) notes that in New Zealand the "time when the household enters the housing market is a crucial determinant of the rate of capital accumulation sustained".

This argument has important theoretical implications. If, as Thorns suggests, "rates of gain are not even across all owners but vary both by location and by the class status of the individual owners", then differences within the owner occupied group can open up between those who gain a lot, those who gain a little, and those who may lose money. This suggests, says Thorns (1981a: 707–8), that rather than owners constituting a single class as Saunders suggests they may comprise "a series of fractions which achieve gain not only at the expense of tenants but also at that of other owner-occupiers". The importance of this point cannot be stated too strongly. The difference between an owner who makes a gain of £200,000, one who makes £2,000 and one who loses £20,000 is considerable. Their attitudes towards home ownership may well be very different, as may their political views and party affliations. The next section examines the distribution of gains and losses by social class, date of purchase, income and other variables in an attempt to identify the relative importance of the factors affecting gains and losses. It shows that while class is an important influence on the size of gains and losses, other factors, particularly date of purchase, are also important.

The Distribution of Gains and Losses by Socio-Economic Group and Income

The following analysis examines the distribution of home owners' gains and losses by socio-economic group of the principal earner. It also looks at other important variables, notably the date of purchase and asks to what extent, if at all, these are class-related. If Saunders is correct that tenure constitutes an independent dimension of social stratification, the distribution of gains and losses should be independent of socio-economic group or social class.

Looking first at what we have termed nominal illusionary gains – that is, the difference between estimated current market value and price paid, there are major differences in mean gains between different socio-economic groups. Based on current property, Table 4.3 shows that professionals had the largest gain

Table 4.3 Mean nominal gains by SEG of head of household

	Illusionary Gain		Crude Gain		
	Current property	First property	Current property	First property	*n*
Professional	63,231	111,780	27,437	73,170	121
Managerial	55,770	106,270	17,681	67,600	230
Other non-manual	41,860	70,520	11,695	39,915	210
Skilled manual	42,420	64,070	19,256	40,861	133
Partly skilled	42,380	50,970	26,300	35,400	67
Unskilled	56,100	60,056	48,611	52,555	18
Mean gain	49,774	84,458	19,315	53,243	779

Source: South East Homeowner's Survey (1993)

(£63,200 on average), followed by managers with £55,800. Leaving aside the unskilled, who comprise a small and statistically unreliable subsample, the other three groups had mean gains of £41,900 (other non-manual), £42,400 (skilled manual) and £42,380 (partly skilled). Gains based on the first property are much greater as the time scale is generally longer but similar differentials were found to exist. Mean gains ranged from £111,800 for professionals and £106,300 for managers, to £70,500 for other non-manual and £51,000 for the partly skilled. The distribution of nominal crude and gross gains conforms to this pattern for the first property, but where current property is concerned, the distribution forms a U-shaped curve with higher mean gains for the skilled and semi-skilled groups than for the managerial and other non-manual groups. The reasons for this are discussed below, but where gains on first property are concerned the absolute differentials by socio-economic group are large and do not support Saunders's thesis. The distribution of mean gains reinforces the inequalities derived from the labour market (Murie, 1991).

The differences reflect several factors. First, higher socio-economic groups buy more expensive properties, which yield greater absolute gains. Nominal illusionary gains on current property ranged from £12,250 on properties with an estimated market value of £56,000 or under, to £132,800 on homes with an estimated market value of £180,000 or over. The pattern of mean nominal illusionary gains by estimated market value is similar for the first property, ranging from £21,200 for properties worth under £56,000 to £227,000 for houses worth over £180,000 (Table 4.4). Professsionals and managers are also more likely than other groups to own detached houses (Table 4.5), and mean nominal absolute illusionary gains vary by property type ranging from £155,000 for detached houses to £25,300 for converted flats.

Table 4.4 Mean nominal illusionary gains by estimated market value (£)

Estimated market value	Current property	First property
<56,000	12,254	21,201
56–69,000	23,123	37,072
70–85,000	30,904	50,177
86–115,000	43,297	68,208
116–179,000	64,713	115,485
>180,000	132,780	226,920
Mean gain	50,480	86,005
n = 961		

Source: South East Homeowner's Survey (1993)

Secondly, absolute gains increase with the number of homes owned. On the basis of first home owned, the illusionary nominal mean gain ranged from £48,167 for those who have owned only one home to £95,880 for those who have owned two homes and £137,646 for those who have owned three or more (Table 4.6).

Table 4.5 Type of property by occupational class (%)

	Professional	Managers	Non-manual	Skilled manual	Semi-skilled	Unskilled
Detached	45	40	19	23	2	2
Semi-detached	17	22	27	35	35	13
Terraced	25	26	35	33	49	58
Purpose-built flats	10	8	12	7	10	16
Conversions	2	3	6	1	4	11
All	100	100	100	100	100	100

Source: South East Homeowner's Survey (1993)

Table 4.6 Illusory gains, first purchase, by number of homes owned

Number of homes owned	Size of gain (£)
1	48,167
2	95,880
3	128,200
4	156,700
5	129,724
6	160,200

Source: South East Homeowner's Survey (1993)

Table 4.7 Mean numbers of years owned by SEG

Professionals	18.1
Managers	16.4
Other non-manual	16.2
Skilled manual	18.2
Semi-skilled	18.1
Unskilled	20.2
Mean all groups	18.1

Source: South East Homeowner's Survey (1993)

The mean number of homes owned also varies by socio-economic group of the principal earner – ranging from a high 2.46 for professionals and 2.33 for managers to a low of 1.37 for the unskilled (see Table 7.8 p. 161).

It is not the case, however, that professionals and managers entered the home ownership market earlier than the other groups. The mean number of years in home ownership varies very little between different groups, averaging around eighteen years for all groups (Table 4.7). It appears that differences in absolute gains on first property by socio-economic group reflect the current price and

Table 4.8 The distribution of real gains (£'s) by SEG

	Illusionary gain		Crude gain		Gross gain
	Current property	First property	Current property	First property	First property
Professionals	23,577	73,887	−12,166	36,210	20,219
Managers	14,985	74,421	−23,058	36,227	16,610
Other NM	16,310	37,770	−14,025	6,712	−16,870
Skilled manual	16,066	31,789	−6,085	7,961	−1,200
Semi-skilled	15,140	23,287	−975	7,671	−5,755
Unskilled	22,939	22,860	−13,123	15,360	2,626
Mean all groups	17,077	51,600	−13,226	20,456	2,768

Source: South East Homeowner's Survey (1993)

type of property owned and the number of moves, rather than length of time in the home ownership market. Professionals and managers have generally traded upmarket more than other groups.

The distribution of real gains on first and current homes (Table 4.8) shows that professionals and managers make much larger absolute gains on the basis of the first home, but on the basis of the current home they make larger crude losses than the skilled and semi-skilled. This reflects the pattern of nominal gains on current property and the date at which different groups purchased their current home. Professionals and managers move more frequently and are thus more likely to have bought near the peak of the 1980s boom. It should be noted that all groups made real crude losses on the basis of their current property and that crude real gains on the basis of first property are much smaller than the equivalent nominal gains. The gross gain measure (including outstanding mortgage and deposit) on first property shows that only professionals and managers made substantial real gross gains.

Higher socio-economic groups tend to earn more and buy more expensive houses and their absolute gains are generally larger. But Saunders (1990) argues that distribution of relative gains is likely to be more or may favour lower socio-economic groups who have tended to buy more recently. What evidence is there for this? Table 4.9 shows that where nominal percentage illusionary gains on first homes are concerned, the professional, managerial and non-manual groups all have higher percentage gains than the manual groups, with professionals and managers having rates of gain considerably above all other groups. But, where current property is concerned the less skilled have higher percentage rates of gain than professional, managerial or other non-manual groups. The differentials persist where nominal percentage crude and gross gains are concerned. Professionals and managers had smaller percentage mean gains than the less skilled on the current home but the differentials were reversed on the first home. The explanation for the higher percentage gains among the less skilled on the current

Table 4.9 Nominal percentage gains by SEG, current and first property

	Illusionary gain (%)		Crude gain (%)	
	Current property	First property	Current property	First property
Professionals	474	2,441	435	2,138
Managers	480	2,173	430	1,894
Other non-manual	603	1,729	570	1,604
Skilled manual	927	1,529	926	1,393
Semi-skilled	811	1,272	769	1,216
Unskilled manual	1,474	1,624	1,439	1,548
Mean all groups	640	1,895	605	1,700

Source: South East Homeowner's Survey (1993)

property is primarily a result of differences in date of purchase between different groups. Because the less skilled bought fewer properties and generally bought their current property earlier than the professional and managerial groups, a lower proportion were caught out by the slump in the market from 1989 and hence they made higher percentage gains on their current property. On the basis of the first property, however, greater long-term gains of professionals and managers are reasserted. Overall, it is clear that the distribution of gains varies with socio-economic group rather than being independent as Saunders asserts.

Gains by Estimated Current Value and Household Income

Table 4.10 shows the distribution of mean absolute nominal gains by current estimated value of property. There is a strong positive relationship with mean

Table 4.10 Mean absolute nominal gains by current value (£)

Property value	Illusionary gain		Crude gain		Gross gain
	Current property	First property	Current property	First property	Current property
<56,000	12,254	21,201	−11,682	−3,488	−17,563
56–69,000	23,123	37,072	−3,795	9,932	−9,623
70–85,000	30,904	50,177	2,636	22,359	−4,678
86–115,000	43,297	68,208	15,930	41,207	8,728
116–179,000	64,713	115,485	30,803	83,189	21,297
>180,000	132,780	226,920	94,478	185,880	83,039
Mean gain	50,480	86,005	19,577	53,647	10,817

Source: South East Homeowner's Survey (1993)

Table 4.11   Mean nominal absolute gains by current household income (£)

Annual household income	Illusionary gain		Crude gain	
	Current property	First property	Current property	First property
<8,150	50,910	64,874	38,145	52,303
8,150–16,249	42,418	56,989	22,467	37,418
16,250–22,499	39,776	58,287	15,398	33,558
22,500–28,899	48,810	77,293	14,936	43,437
28,900–39,999	52,863	102,116	8,398	57,663
>40,000	55,068	130,413	5,402	76,393
Mean gain	48,772	82,769	17,735	50,543

Source: South East Homeowner's Survey (1993)

illusionary gains on current property ranging from £12,300 on houses worth under £56,000 to £132,800 on houses worth over £180,000. Measured on first property, the gains range from £21,200 to £226,900. For crude and gross gains households owning cheaper property made losses, with lower-priced houses having the greatest losses. This is not surprising as cheaper properties are likely to have been recently bought by first-time buyers who have little opportunity to build up their gains during previous years. They had no cushion against the fall in prices since 1989. In general, the more valuable the property, the greater the size of nominal gains.

The analysis of gain distribution by income is more complex than that for other variables, primarily because of the differences in income distribution between those households where the principal earner is economically active and those where the principal earner is retired. The latter are generally older households who have owned for longer and have made substantial gains on the basis of their first property, even though their incomes are currently relatively low. The distribution of mean gains on current property by income is U-shaped. The groups gaining most are those with a household income of under £8,150 (generally older, retired owners living on a pension) and those earning over £28,900, who live in more expensive properties. The distribution of mean gains on first property is more linear, gains rising steadily with mean income with the exception of those with mean incomes of under £8,150 (the retired) who have higher gains than the next two income groups (Table 4.11).

This is clearly shown in the distribution of incomes in households where the principal earner (PE) is economically active and those where the PE is retired. Two-thirds (65.4%) of households where the PE was retired had an annual income of less than £10,625 compared to just 16% of households where the PE was economically active. The distribution of illusionary gains on current property broken down by economically active and retired principal earners is linear, with the higher income groups making the largest gains.

Variation in Gains by Date of Purchase

The consistent variations in the pattern of gains/losses by socio-economic group when based on first home compared to the current home point to the key role of date of purchase as a determinant of gains. Although all socio-economic groups have owned for a similar length of time on average there are marked differences in the length of residence in the current home by socio-economic group. Because professionals and managers have owned more homes and trade up more frequently, they have generally owned their current home for a much shorter time than the less skilled. As prices in the South East rose unevenly until 1989, it would be expected, other things being equal, that the longer an owner has owned, the greater the size of absolute gains, with considerable boosts if s/he bought before or at the start of one of the three major house price booms (1971–73, 1977–79, 1983–89). As prices have fallen sharply since 1989, by some 30% in London and the South East, the reverse would be expected for those individuals who bought or moved upmarket at or near the peak of the late 1980s boom.

Looking at the nominal illusionary gain on current home, those who bought before 1959 had average gains of £106,800. This fell to £86,100 for those who bought in the period 1960–68, to £62,700 for those who bought in 1969–73 and to £30,100 for those who bought in 1980–84. For those who bought in the two most recent periods, the mean gain was much smaller, £8,200 for those who bought in 1985–88, and just £360 for those who bought after 1988. This highlights the very different experiences of those who bought in the 1980s boom or the subsequent slump. They gained very little on average and a high proportion have made losses. The distribution of gains measured on the basis of first property is similar, though gains are much higher. They range from £137,000 for those buying in 1960–68 to just £850 for those buying in 1989–92. Similarly for crude gains, although recent buyers made large losses, averaging £52,000 for those who bought since 1988. Blessed are those who buy early when prices are rising (Table 4.12 and Figure 4.5).

It is clear that year of purchase is one of the most important determinants of gains and losses from the home ownership market. Until the early 1990s the longer the time since purchase the greater the price inflation and the greater the gains. But for those who bought after 1980, particularly for those who bought after 1985, the link between date of purchase and gains has become negative: the more recent the purchase the greater the likely losses (Figure 4.5). Gains are also related to the age of principal earner, which is an indirect measure of date of purchase. Table 4.13 shows that the older the owner the greater the gains, up to a peak of 60–64. The table also shows that the most recent cohort of owners, those aged 20–29, have made minimal gains from ownership measured on the basis of their current property. Whether, and when, they will eventually recover from the impact of the slump in house prices remains to be seen.

Table 4.12 Mean nominal illusionary gains by year of first purchase (current and first property)

	Illusionary gain (£)		Crude gain (£)		
	Current property	First property	Current property	First property	*n*
Pre-1959	106,800	130,261	102,888	126,495	109
1960–68	86,130	137,292	75,404	122,504	109
1969–73	62,730	113,213	38,857	83,898	103
1974–76	60,170	119,288	29,117	84,433	65
1977–79	62,645	102,484	27,684	69,193	62
1980–84	30,100	73,470	–13,599	29,190	148
1985–88	8,230	26,386	–40,328	–21,385	132
1989–92	360	851	–52,240	–51,734	81
Mean gain	50,500	86,160	19,513	53,717	809

Source: South East Homeowner's Survey (1993)

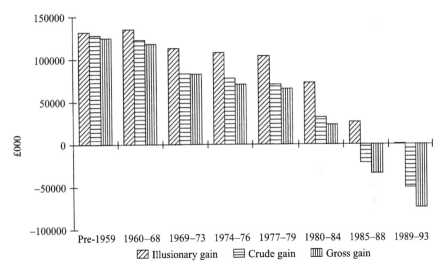

Figure 4.5 Comparison of gain measures by year of purchase, first property
Source: South East Homeowner's Survey (1993)

The Importance of Area and Current House Type

There were also marked differences in gains between the local authorities surveyed. The nominal illusionary mean gains on first property ranged from £142,300 in Chiltern UDC (an area characterised by large, detached houses and residents at a later stage of their housing careers) to £57,100 in Milton Keynes (Table 4.14).

Table 4.13 Mean illusionary gains by age of principal earner

Age	Mean gain (£)		
	Current property	First property	*n*
20–29	640	8,153	72
30–39	15,843	54,593	204
40–49	56,020	110,205	196
50–59	64,500	101,928	137
60–64	101,985	126,271	68
65–74	78,500	105,337	82
75+	89,333	99,365	51
All ages	50,586	86,053	812

Source: South East Homeowner's Survey (1993)

Table 4.14 Nominal absolute mean gains by area (£)

	Illusionary gain		Crude gain		Gross gain	
	First property	Current property	First property	Current property	Current property	*n*
Chiltern	142,317	76,100	104,270	42,927	35,200	186
Hammersmith	95,150	66,650	62,540	35,235	15,200	107
Haringey	68,439	50,440	37,040	17,900	8,600	173
Oxford	70,725	48,000	50,700	28,942	24,400	160
Milton Keynes	57,150	20,900	22,280	−13,912	−20,260	201
Mean gain	86,000	50,480	53,650	19,577	10,800	824

Source: South East Homeowner's Survey (1993)

The differentials were more marked for crude gains, ranging from £104,300 in Chiltern to £22,300 in Milton Keynes. This reflects the fact that owners in Chiltern had generally owned for longer and paid off a larger proportion of their mortgage than owners in Milton Keynes and Haringey, where owners tend to be younger and have bought more recently. Mean number of years of home ownership ranged from 22 years in Chiltern to 14.7 years in Haringey and Milton Keynes.

Not surprisingly, current house type and current value are also very strongly related to mean gains. The absolute illusionary gain on first property ranges from £155,200 for detached houses to £25,000 for converted flats. This, of course, is partly a reflection of the lower price of flats. Crude and gross gain measures show losses for flat owners, and these losses are more marked if measured on the basis of current property (Table 4.15). This reflects the fact that most flat owners have bought more recently and are at an earlier stage in their home ownership careers.

Table 4.15 Mean absolute gains (£'s) by current property type

	First property			Current property	
	Illusionary	Crude	Gross	Crude	Gross
Detached	155,199	114,055	106,650	40,071	32,766
Semi-detached	72,704	44,495	38,059	20,230	13,289
Terraced	64,924	38,059	27,814	17,418	7,862
Purpose-built flat	28,833	−5,462	−16,045	−17,776	−28,394
Converted flat	25,322	−4,733	−22,964	−12,167	−30,357

Source: South East Homeowner's Survey (1993)

The Relative Importance of Social Class, Purchase Date and other Variables in the Determination of Mean Gains

The undoubted importance of year of first purchase as a determinant of gains and losses raises the question of the relationship between date of purchase and social class. Put simply, does the influence of year of first purchase outweigh the influence of class? While the gains of a professional or manager who bought in the mid or late 1980s are likely to be far less than those of a skilled manual worker who bought in the 1970s and early 1980s, a professional or manager who bought an expensive house in the 1960s or 1970s may also be expected to have greater absolute gains than a skilled worker who bought a cheaper house at the same time.

To shed light on this issue, Table 4.16 and Figures 4.6 and 4.7 shown illusionary mean gains by socio-economic group by date of purchase by first and current property grouped into time periods. They show that on the basis of first property, professionals and managers generally gain more in absolute terms in every period than other groups, and the gains fall systematically with socio-economic group. Based on current property, higher socio-economic groups generally gain more for property purchased before the early 1980s, when the pattern breaks down. For current houses bought since 1980, professionals and managers do not gain more than other groups. The table also shows that on either first or current property the less skilled who bought in the 1960s or 1970s or before make larger gains than the highly skilled who bought in the 1980s. Purchase date seems to outweigh the effects of class over the long term, although in any given period, the higher social classes gained more where first property is concerned. For those who bought before 1959, the mean gain for professionals was £170,400 falling to £75,400 for the semi-skilled. For those who bought a first property in 1985–88, professionals gained £45,400 on average compared to £18,200 for semi-skilled and £13,857 for skilled manual workers (Figure 4.7). Holding purchase date constant, the higher social classes gain more in each period

Table 4.16 Variations in mean absolute illusionary gains by SEG of major earner and time of first purchase

Date of purchase	Prof.	Man.	ONM	Skilled	Semi
First property					
pre-1959	170,353	187,038	113,760	78,375	75,400
1960–68	162,947	198,464	110,556	91,353	74,692
1969–73	146,286	143,094	101,286	73,182	70,167
1974–76	112,556	115,000	136,067	88,538	59,250
1977–79	122,909	110,353	95,143	99,778	88,250
1980–84	104,476	83,326	56,250	64,346	44,444
1985–88	45,400	30,975	20,481	13,857	18,200
1989–92	−3,111	5,250	−2,242	1,286	1,000
Current property					
pre-1959	148,147	142,160	96,250	66,687	74,700
1960–68	107,737	91,407	82,593	75,000	51,750
1969–73	61,000	62,500	65,143	49,636	53,500
1974–76	54,444	68,947	47,467	57,308	43,000
1977–79	91,727	67,875	51,550	41,444	–
1980–84	33,857	34,043	24,025	33,840	42,667
1985–88	750	12,000	3,111	6,762	15,533
1989–92	−3,111	5,250	−2,242	1,286	−5,667

Source: South East Homeowner's Survey (1993)

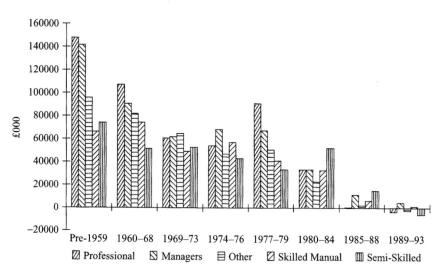

Figure 4.6 Nominal illusionary gains by year of purchase and SEG, current property
Source: South East Homeowner's Survey (1993)

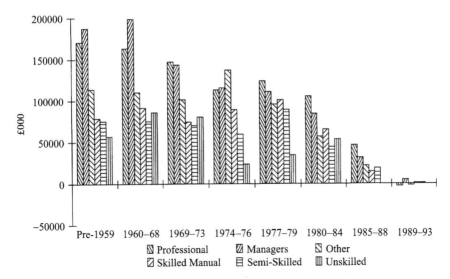

Figure 4.7   Nominal illusionary gains by year of purchase and SEG, first property
Source: South East Homeowner's Survey (1993)

measured on the basis of first property and in each period until the early to mid 1980s measured on current property. The influences of class and date of purchase on mean gains therefore interact in a straightforward way until the early 1980s.

The year of first purchase is one of the most important determinants of gains and losses from the home ownership market: the scale of variation outweighs that of class, income, age or area. But this is not very surprising. The level of house price inflation from the late 1960s to the late 1980s has been marked, as have been the price falls since the late 1980s: gains and losses are quite strongly time-dependent phenomena and are closely related to date of purchase and subsequent price movements.

The Relative Importance of Different Influences on Determination of Gains

Clearly, a number of different factors influence the size of absolute gains: social class, date of first purchase, the number of homes owned, age, current property type and value. There are also strong interrelations between these factors. To shed some light on the statistical importance of various factors some multiple regression analyses were undertaken with absolute illusionary gains as the dependent variable. Regression analysis is a statistical tool which enables us to determine the extent to which variations in a variable we are interested in, in this case illusionary gains, can be statistically explained or predicted by a number of independent variables. Data are available for each variable for each household and the independent variables included were:

1. Indexed original cost of first property (FHV)
2. Year of First Purchase (YFP)
3. SEG of principal earner (SEGPE)
4. Area (AREA)
5. Type of current home (TYPE)
6. Number of homes owned (NUMBR)
7. Combined current household income (TOTINC)

The matrix of correlation coefficients between variables is shown below. It can be seen from the matrix that type and number of homes owned are quite strongly related (−0.354), as are total household income and socio-economic group of the principal earner (−0.425), type and year of first purchase (0.297). But the strongest correlations are between gains and year of first purchase (−0.468), gain and type (−0.454), gain and number of properties owned (0.399), and gain and area (−0.347). This is reflected in the regression, which had a multiple R of 0.732 and an $R2$ value of 0.536 where the maximum value is 1.0. Put in less statistical language, the independent variables explain 53% of the total variation in absolute illusionary gains between households. The regression equation is given below with standardised B coefficients for each variable. The SEG of the principal earner, area and house type are dummy variables rather than continous variables. For example, a detached house was allocated the value 1, a semi-detached house value 2, a terraced house 3 and so on. So too, the socio-economic group of a professional was given the value 1, a manager 2, and so on.

Correlation matrix for first property

Gain	FP	FHV	YFP	SEGPE	AREA	TYPE	NUMBR	TOTINC
	1.000	−.101	−.468	−.258	−.347	−.454	.399	.319
FHV		1.000	−.011	−.046	−.074	.103	−.036	.107
YFP			1.000	−.018	.126	.297	−.313	.175
SEGPE				1.000	.159	.137	−.199	−.425
AREA					1.000	−.053	−.017	−.173
TYPE						1.000	−.354	−.165
NUMBR							1.000	.247
TOTINC								1.000

1. Regression Equation for Absolute Illusionary Gain on First Property

Illus FP = 331.9 − 0.125 FHV − 0.372 YFP − 0.06 SEGPE − 0.266 AREA − 0.256 TYPE + 0.18 NUMBR + 0.259 TOTINC ($R = 0.732$, $R2 = 0.536$)

It is clear from this that the most important single variable to emerge was year of first purchase with standardised beta coefficient of 0.372. All the variables were significant at the 0.001 level with the exception of SEG (0.004). The result of the regression shows that the most important determinant of size of absolute gains over time is the year of purchase (0.372), followed by area (0.266), income (0.259), and type of current property (0.256), with SEG having only a minor effect (0.06).

The results of an identical regression equation run on the current property is shown below. It gave a lower multiple R of 0.640 and an $R2$ of 0.410. The correlation coefficients with gains differed considerably, however (below). The correlation with number of homes owned was insignificant, and the coefficient with type of home and socio-economic group was greatly reduced, while correlation with year of first purchase rose to -0.499. The change in correlation coefficients was reflected in the beta coefficients. The coefficient for year of purchase rose to 0.507. This was followed by area (0.260), number of homes owned (0.241) and property type (0.225). Socio-economic group was again of minimal importance with a beta coefficient of 0.096. These changes are not surprising, as the house price slump means that timing of purchase is more important in determining gains and losses over time on current than first property.

Correlation coefficients with gains on current property

Gain	CP	CHV	YFP	SEGPE	AREA	TYPE	NUMBR	TOTINC
	1.000	.020	−.499	−.134	−.303	−.272	.002	.051

2. Regression Equation for Absolute Illusionary Gain on Current Property

Illus CP = 347.2 − 0.143 CHV − 0.507 YFP − 0.096 SEGPE − 0.260 AREA − 0.225 TYPE + 0.241 NUMBR + 0.137 TOTINC (R = 0.640, $R2$ = 0.41)

Conclusions

What can we conclude from this analysis of the distribution of gains? First, illusionary gains are greater than crude or gross gain, which follows logically from the construction of these measures. Secondly, and unsurprisingly, as we move from illusionary to crude to gross measures the proportion of losers increases sharply, which reflects the inclusion of mortgage costs, deposits and the like. For the UK as a whole in 1991, the proportion of losers on the different measures for current property rose from 5% (illusionary) to 40% (crude) and 47% (gross). In the South East in 1993 the proportion of losers measured on first property rises from 8% to 28% to 31%, and on the basis of current property the proportion of losers rises from 20% to 44% to 46%. If the measures are calculated in real rather than nominal terms the proportion of losers is even greater, rising to 58% of owners on current property. These results completely undermine Saunders's claims that most owners gain. This may have been true in nominal terms until 1989 but not after.

Thirdly, higher socio-economic groups made higher gains than other groups when measured on the basis of first property. The pattern of gains on the current property was U-shaped, reflecting the fact that professional and managerial owners tend to move more frequently and had moved upmarket in the 1980s, thereby exposing themselves to price falls. Measured over their entire housing career, however, professional and managerial owners gained almost twice as much in absolute terms as manual groups, which reflects the more expensive property they are able to buy. The distribution of percentage gains is more complex. Measured on the basis of current property, the higher socio-economic groups had much smaller percentage gains than the less skilled, but on the basis of first property (i.e. over their housing career) the pattern was reversed. In the long run, professionals and managers gain more than other groups in both absolute and relative terms. The price falls of the 1990s disrupted this pattern, however, and the less skilled, who move less frequently, gained more. The distribution of gains by income is complex, but this is primarily a result of aggregating older, retired, outright owners with the younger, mortgaged owners on higher incomes. Excluding the former, gains are directly related to household income: the higher the income, the greater the mean gains.

The most marked variation in absolute gain is between mean gains and year of first purchase or length of time in the home ownership market. The longer one has owned, the greater the gains, whether measured on the basis of first or current property. This is not surprising, and reflects the long-term trend of upward price appreciation. There were considerable variations in gains by area and current house type, but the variations by area largely reflect the variations in house type and stage in the housing career, the largest absolute mean gains being made on larger, more expensive detached houses and the biggest losses on flats, which tend to be bought by young, first-time buyers.

When we look at the relationship between social class and date of purchase, it is clear that the higher social classes made the largest gains in every time period on first property until 1989 and on current property until the end of the 1970s. Had the slump of the 1990s not taken place we can suggest that the association between size of gains and class would have held for every time period. The longer one had owned, the greater the gains, professionals and managers gaining more in each time period, with long-term less skilled manual owners making more than short-term professional owners. But when we compare like with like, that is members of different classes who bought at the same time, the higher social classes gain more because they generally live in more expensive properties. Class and income strongly influence gains for comparable cohorts of buyers, but over the long term date of purchase is the most important determinant of absolute gains. An unskilled worker who bought in the 1960s or 1970s will, almost inevitably, have a larger gain than a professional or managerial owner who bought in the last few years but, when, length of time in the housing market is held constant, social class reasserts its importance.

CHAPTER 5

Who Gets What? The Distribution of Home Owners' Equity in Britain

Introduction

It may seem strange to talk about scale and distribution of home owners' equity in the aftermath of the longest and deepest slump in the home ownership market in Britain for many years. But, despite the marked falls in nominal house prices in Britain since 1989, and the even sharper fall in real prices (taking inflation into account), the surge in repossessions, and the rise of what came to be known as "negative equity", the great majority of home owners in Britain, and all outright owners, continued to have substantial positive equity in their houses totalling several hundred billion pounds throughout the slump. Not surprisingly, media attention focused on those owners who suffered in the slump, but this should not distract us from the fact that the growth of home ownership in Britain since the War along with the massive house price inflation during the 1970s and 1980s has been associated with a major redistribution of wealth. As I shall show, housing wealth is one of the most equally distributed forms of wealth ownership in Britain, far more equal than ownership of land or stocks and shares and other traditional forms of wealth holding. This is not to say that housing wealth is evenly distributed, but that it is far less unequally distributed than other forms of wealth holding. Sixteen million households in Britain now own or are buying a house, and for most of them it is their largest capital asset. This chapter explores the importance of housing wealth and the distribution of housing equity, both positive and negative.

Home Owners' Equity and Housing Wealth in Britain

The growing importance of housing wealth in Britain results from two main factors. The first is the growth of home ownership since the War, and the second is the rapid increase in house prices during the 1960s, 1970s and 1980s. As Chapter 3 pointed out, the level of home ownership in Britain has increased rapidly during the past 40 years. In 1950 there were some 4 million owner occupied dwellings, 29% of the total. By 1996, there were 16 million owner occupied dwellings, 68% of the total (Housing and Construction Statistics, CML, 1997). The thirty years from 1960 to 1989 also saw rapid house price inflation over and above the general level of inflation. National average house prices rose

Table 5.1 The importance of different assets in gross and net personal wealth, 1994 (% of total assets)

	Gross	Net
UK residential buildings	45.2	40.0
Insurance policies	19.4	17.1
Cash, including bank and interest-bearing accounts	17.1	15.1
Stocks and shares	12.7	11.2
Loans and mortgages	3.7	3.2
Other personalty	3.2	2.8
Household goods	3.7	3.3
UK land and buildings	2.9	2.6
Government securities	2.4	2.1
Trade assets and partnerships	2.2	2.0
Foreign immovables	0.6	0.6
Total	113.1	100
Less: Mortgages	−8.6	
Other debts	−4.5	

Source: Inland Revenue Statistics, 1997 Table 13.1

from £2,500 in 1960 to £5,000 in 1970 and £65,000 in 1990 (CML, 1997). This represents an increase of 1,200% in nominal terms in just thirty years, and in real terms the value of housing rose about threefold (Holmans, 1990). These trends have, between them, led to great changes in the importance of property ownership in personal wealth and in inheritance. Not only do many home owners now own a substantial asset, but a large proportion of net personal wealth and inheritance consists of housing (Murie and Forrest, 1980; Munro, 1989; Hamnett, 1995b and c; Forrest and Murie, 1989).

The Royal Commission on the Distribution of Income and Wealth (1977) estimated that the value of dwellings as a proportion of net personal wealth increased from 18% in 1960 to 37% in 1975, and this had risen to just over 50% by 1990. The proportion has subsequently fallen back as a result of the sharp slump in the home ownership market – prices in London and the South East fell by 25% in cash terms between 1989 and mid 1995, and by far more in real terms. In addition, the period from mid 1992 to 1997 saw a sharp rise in share prices, which will have increased the relative importance of shares in net personal wealth. But the 1996 Inland Revenue statistics show that residential buildings accounted for 45.7% of net identified personal wealth in 1993. This is a fall of six percentage points from the 51.6% of 1989, and parallels the figure of 46.8% in 1987, prior to the peak of the last house price boom. Nonetheless, ownership of residential property now accounts for over 45% of personal wealth in Britain, far above other assets such as stocks and shares (Table 5.1). Because the data (Series C Marketable Wealth) are based on figures derived from estates passing at death, they exclude pension rights, but insurance policies figure strongly. This would not be the case if the figures were based on the living population.

Housing is more than twice as important as insurance policies, almost three times bank deposits and almost four times stocks and shares. The significance of this transformation cannot be overstated.

Research on the changing asset composition of estates confirms the work of the Royal Commission on the Distribution of Income and Wealth. In 1904 house and business property accounted for 16% of the total gross capital value of estate, falling slightly to about 14% in the interwar period. In 1968–69 house property accounted for 23% and by 1986–87 it had almost doubled to 41%. If the share of house property in estates rises, something has to fall, and the principal losers were government securities and stocks and shares, which fell from a peak of 55% in 1929–30 to 34% in 1968–69 and to just 21% in 1986–87. The other main gainer was cash, bank and other deposits, which doubled from 12% in 1949–50 to 24% in 1968–69. This increase in housing wealth in the 1970s and 1980s is of great importance. The value of housing wealth rose steadily from £13 billion in 1957, when it comprised 23% of net personal wealth, to £82 billion in 1972, doubling to £166 billion in 1977, to £322 billion in 1980 and again to £663 billion in 1990, when it comprised 52% of net wealth (Pannell, 1992). The asset composition of personal wealth holdings was transformed with important implications for consumer spending and economic management, as Chapter 8 shows.

Home Ownership and the Reduction of Wealth Inequality

The growing importance of housing wealth within personal wealth holdings had significant implications for the degree of wealth inequality in Britain, which fell sharply from the 1920s until the mid 1980s. According to Atkinson et al. (1989), in 1923 the top 1% of wealth holders owned 61% of wealth in England and Wales, the top 5% owned 82%, and the top 10% owned 89%. By 1938, the share of the top 5% had fallen slightly, but no major change had occurred. By 1960, however, a dramatic transformation had taken place. The share of the top 1% of wealth owners had fallen to 34% (down 21 points from 1938), the share of the top 5% had fallen to 60%, and the top 10% had fallen to 72%. This transformation continued at a slower pace until 1981, when the share of the top 1% had fallen to 23%, the share of the top 5% to 46% and the top 10% to 63% (Table 5.2).

What is remarkable about these figures is that, while the share of the top 1% of wealth owners fell dramatically, the share of the top 20% fell relatively little: from 94% to 82%. This points to a major redistribution within the top 20%. Indeed, Atkinson et al. (1989: 319) point to "the relative constancy of the share of the 4 percent of top wealth-holders immediately below the top 1 percent. In 1923, this group owned 21.1 percent of the personal wealth in England and Wales; in 1981 the equivalent figure was 23.2 percent." Atkinson (1983) suggests four possible reasons for the reduction in wealth inequality since the War: first, the role of higher tax rates on large estates and the consequent incentive for

Table 5.2 Percentage shares in the distribution of personal wealth in Great Britain

	top 1%	5%	10%	20%	25%	50%	
1923	60.9	82.0	89.1	94.2	–	–	(England and Wales)
1938	55.0	77.2	85.4	91.6	–	–	GB
1960	34.4	60.0	72.1	83.6	–	–	GB
1970	30.1	54.3	69.4	84.9	86.0	97.0	GB
1981	22.5	46.0	62.8	82.5	75.0	84.0	GB
1985	20.0	40.0	54.0	n/a	76.0	93.0	GB

Source: Atkinson et al. (1989), Central Statistical Office (for figures since 1960)

tax avoidance, particularly gifts of wealth prior to death; secondly the increase in wealth holding among women, particularly the tendency for married couples to hold wealth jointly; thirdly the changes in the relative price of different assets held by different wealth groups: a very high proportion of personally held shares are owned by the top 5% of wealth holders. The final factor is the growth of owner occupation. As he put it (1983: 171): "One of the most striking social changes since the beginning of this century has been the increase in owner occupation. Between 1900 and 1970, the proportion of owner occupied dwellings rose from around 10% to 50%. Coupled with the rise in house prices, this must have had a profound effect on the (wealth) distribution."

Conversely the sharp decline in private landlordism has also reduced the concentration of house ownership. In 1914, 90% of dwellings were privately rented, but by 1991 this figure had fallen to 8%. As Revell (1967: 381) noted: "when one individual owned several houses . . . let out as an investment, he had a good chance of appearing in the top 1%, whereas the houses now appear in a number of smaller estates". The decline of private landlordism and the growth of home ownership can therefore be seen as two sides of the same coin in terms of housing wealth, which has become both far more important and more widely spread over the population as a whole. Home ownership represents the first truly mass asset.

Atkinson et al. (1989) devised a mathematical model to predict the relative share of the top 1% at different dates, and this suggested that between 1972 and 1981 a very large part in the reduction of the share of the top 1% could be attributed to the expansion of popular wealth as a result of the spread of owner occupation and the rise in house prices. More generally, they suggest that the relative share of the top 1% of wealth owners owes a great deal to the relative impact of house price inflation and rises in share prices. When the value of shares rises more rapidly than house prices, the share of the top 1% increases. A similar conclusion was reached by the Royal Commission on Distribution of Income and Wealth: "The considerable growth in the net stock of dwellings combined with the large increase in price had the effect of reducing the share of the top one percent of wealth holders by 6 percentage points between 1960 and 1972."

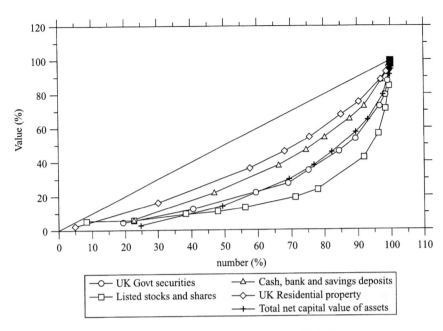

Figure 5.1   The distribution of different assets among wealth holders

Source: Inland Revenue Statistics (1990)

Housing wealth is not just widely owned, it is the most equally distributed of all assets. The degree of inequality of different forms of wealth holdings is reflected in Figure 5.1, which shows the cumulative distribution of different assets across the population. The horizontal axis is the cumulative percentage of population from 0 to 100, and the vertical axis is the cumulative percentage of wealth in a given asset. The 45% line represents a perfectly equal distribution where the bottom 1% of the population owns 1% of wealth, the bottom 10%, 10%, the bottom 50% owns 50% and so on. All wealth distributions differ from this perfectly equal distribution, and the flatter the curve, the more unequal the distribution. Figure 5.1 shows that residential property is the most equally distributed of assets among wealth holders, and stocks and shares the most unequally distributed. These distributions exclude those who own no wealth. In the case of housing this will exclude tenants and those owners who have no equity in their home. Indeed, because the figures are based on assets passing at death and are thus based on outright owners, they overstate the housing wealth of owners as a whole, particularly those in an early stage of their housing careers. It should also be pointed out that the degree of wealth inequality has risen again since the mid 1980s (Good, 1990), partly as a result of Thatcherite economic and fiscal policies which have favoured the wealthy and partly as a result of the rising stock market. Nonetheless, as Feinstein (1992) points out, a revolution has taken place in the distribution of wealth since the War, and home

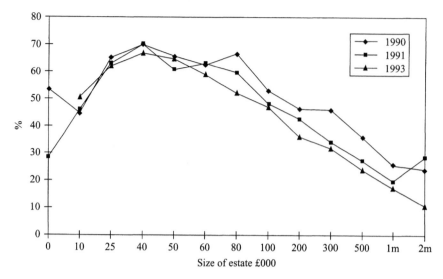

Figure 5.2 Residential property as a percentage of identified personal wealth, 1990, 1991, 1993

Source: Inland Revenue Statistics (various years)

ownership has played a key part in this. Before the War a small proportion of the population owned all the assets worth speaking of, whereas now a substantial proportion own or are buying their own homes. This is a change of great significance.

The Distribution of Housing within Personal Wealth Holdings

Although housing is one of the most evenly distributed assets in personal wealth it is not evenly distributed across wealth bands. Not surprisingly, it is most important in the middle wealth bands from £25,000 to £60,000, where it accounts for well over 60% of identified personal wealth. Below £25,000 housing is relatively unimportant compared to cash and interest-bearing deposits, and over £200,000 other assets such as stocks and shares become steadily more important. In 1993, the latest date for which Inland Revenue figures are available, housing accounts for over 60% of identified personal wealth in the middle wealth bands, rising to 67% in the £40,000–50,000 band. Even in the £100,000– 200,000 band it still accounts for 46% of personal wealth (Figures 5.2 and 5.3). The importance of housing in net wealth has fallen from 1990, particularly in the wealth bands over £200,000, but this is expected given the housing slump and the rise in share prices. If house prices continue to rise over the next two to three years and if share prices experience a downward correction, the importance of housing in personal wealth is likely to increase again.

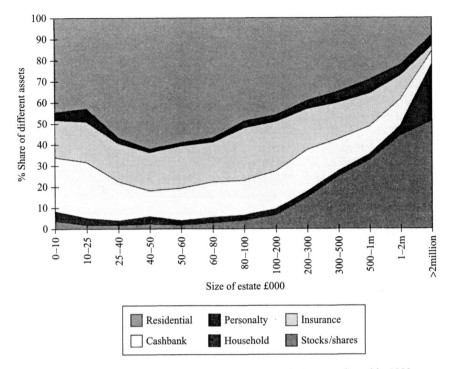

Figure 5.3 The percentage distribution of assets in identified personal wealth, 1993
Source: Inland Revenue Statistics (1996)

The BHPS Data on Housing Wealth

The Inland Revenue statistics on estates passing at death and on personal wealth holdings provide the best data on overall personal wealth distribution currently available. But, not only do the wealth figures have the disadvantage of being derived from a sample of the dying population, with the potential biases that entails, they also contain no information on the breakdown of wealth holdings by age, region, household type, region or income. In order to get a detailed picture of the distribution of housing wealth in Britain we need a source of survey data which contains both housing wealth and other data.

The solution was provided by the British Household Panel Study. This is an annual panel survey of some 5,500 households in Great Britain organised by the University of Essex and begun in 1991. It contains a section on housing and housing tenure and, for owners, original price paid, mortgage, estimated current value and other information as well as individual and household income. Inevitably, as it is not purpose-designed, it presents certain problems, but it offers the best available current source of information on housing wealth. The 1991 data are used here. This is not a problem, not least because the home ownership

107

market has been depressed since the late 1980s and has not experienced significant price inflation until 1996. The BHPS includes 3,672 (66.6%) owners (of whom 35% owned outright and 65% were buying on a mortgage). This parallels figures in the 1991 Census and the General Household Survey.

The best definition of net housing wealth is the current market value of all house property owned by the household less any outstanding mortgage debt or other loans secured on the home. Ideally this should be calculated using actual sale price or current estate agent valuations, and data on outstanding mortgages, if any. Unfortunately, sale values or valuations are rarely available, and the BHPS asked respondent owners to estimate current market value of this "home". This poses certain problems, and the BHPS did not ask a question on the amount of outstanding mortgage, which had to be calculated using data from other BHPS questions. We consider, however, that the estimates we derived are reasonably robust.

Distribution of Housing Wealth

Distribution of Current House Values among Home Owners

The distribution of housing wealth among home owners is examined in the following section. First, however, it is useful to look at the distribution of current house values as this gives an indication of the range of properties in which people live. Two methods of estimation were employed – home owners' estimates and a check using purchase prices inflated by DoE price data. The two distributions are shown in Figure 5.4. The median category for both estimates was identical at £50,000–61,000 and mean estimated value was £79,300 against £80,900 for the DoE adjusted purchase price. These means are higher than building society purchase prices but this is to be expected as building society prices are based on current purchases which include a large number of first-time buyers and exclude outright owners who tend to live in more expensive houses. Figure 5.4 shows that 62% of households live in properties with an estimated value of between £38,000 and £85,000, 26% in properties with an estimated value of over £85,000, and only 6% in properties valued over £170,000. At the other end of the scale, 12% live in homes with a value of £37,000 or less. These are likely to be small terraced houses or flats in the North of England. The 0.7% of values of under £13,000 are caravans and other similar dwellings.

Distribution of Housing Wealth among All Home Owners

The distribution of housing wealth among home owners in 1991 is shown in Figure 5.5. Mean household housing wealth was £60,900, but there was a broadly spread negatively skewed distribution, with 11% of owners having under £14,000, 26% having £14,000–38,000, 28% having £38,000–61,000 and 16%

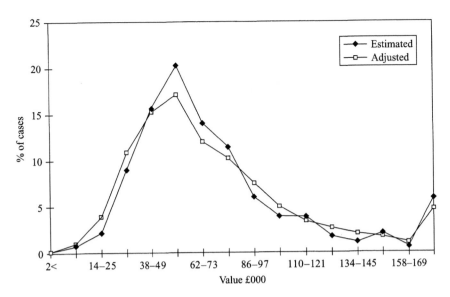

Figure 5.4   Estimated and adjusted value of houses, 1991
Source: British Household Panel Survey (1991)

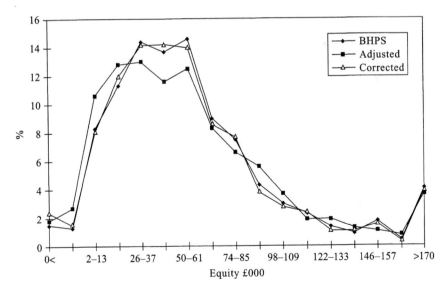

Figure 5.5   Distribution of home owners' equity, 1991
Source: British Household Panel Survey (1991)

£62,000–85,000. There is then a long tail with the remaining 18% widely spread, 11% having over £110,000 and 4% over £170,000. Tenants have no housing wealth. If they are included, the mean household housing wealth in Britain across all tenures in 1991 was £40,200. These figures show that most home owners had significant housing equity, notwithstanding the slump in house prices from 1989 to 1991. If the BHPS data for subsequent years were used these could show a distinct downwards shift as house prices fell.

Distribution of Housing Wealth among Outright and Mortgaged Owners

Not all owners had positive housing wealth. Using the estimated values, 1.5% of mortgagors had negative equity with a mean of –£7,800 and a median of –£4,450. The figure of 1.5% gives a grossed up national total of 225,000 home owners with negative equity – well below estimates which suggested a figure of 1 to 1.4 million owners in 1991 (Wrigglesworth, 1992; Bank of England, 1992; Gentle et al., 1994). This may be a reflection of the methodology employed, with some mortgagors overestimating the value of their homes, thus leading to an underestimation of negative equity.

One solution to the problem of household overestimation by mortgagors is to deflate their estimated equity by 3% to bring it into line with the adjusted estimates. This gives a figure of 2.4% of all owners with negative equity: a grossed up national total of 370,000. Not surprisingly, given the spatial and temporal character of the slump, 59% of those with negative equity were found in London and the South East, 89% had bought since 1988 and 57% were aged 20–29. A specific cohort of younger, southern buyers, who bought in the mid to late 1980s when prices were at their highest, suffer from negative equity, but recent price rises in the South will have reduced this to some extent.

Variations in Mean Home Owners' Equity

The distribution of home owners' equity varies considerably according to region, socio-economic group, age and so on. The more important of these are examined below, including the distinction between mortgaged and outright owners, with outright owners tending to be older and on lower incomes but with higher equity. Because of the regional variations in house prices there is a marked regional difference in equity, with average estimated equity in London, the South East and the South West of £75,000, falling to £57,000 in the East and West Midlands, £49,000 in Yorkshire and a low of £33,000 in Scotland. Adjusted purchase price equity ranged from £87,600 in East Anglia and £78,000 in London and the rest of the South East (ROSE) to £56,000 in the Midlands and £28,900 in Scotland (Table 5.3). The differences in results between the methods are not important. They both show broadly the same pattern with mean equity higher in the South and lower equity in Scotland, the Midlands and the North.

Table 5.3 Mean equity (estimated and adjusted) of outright and mortgaged owners by region (£)

Region	Owner's estimated value-based			Adjusted purchase price		
	All owners	Outright	Mortgaged	All owners	Outright	Mortgaged
London	75,000	102,200	61,700	78,300	107,100	66,600
ROSE	76,700	102,400	64,400	79,000	121,100	61,100
SW	75,200	106,200	50,300	67,200	97,700	46,200
E Anglia	69,900	91,900	52,600	87,600	139,900	51,900
E Mid.	56,900	69,200	49,600	55,300	75,000	44,400
W Mid.	56,800	80,800	45,100	56,400	94,300	38,800
NW	54,400	65,200	48,300	47,100	66,200	38,000
Yorks/H	49,200	59,500	42,800	49,000	74,800	35,300
North	44,200	63,500	34,700	35,200	57,300	26,400
Wales	51,500	72,400	34,100	49,500	83,200	28,800
Scotland	33,400	50,900	27,200	28,900	58,900	20,700
GB mean	61,000	82,200	49,100	60,300	92,600	44,900

Source: British Household Panel Survey (1991)

The table also shows variations in mean equity for outright and mortgaged owners. Where outright owners are concerned, owners' estimated mean equity varies from a high of just over £100,000 in London, the rest of the South East and the South West to a low of £51,000 in Scotland. For mortgaged owners, estimated mean equity varied from a high of £64,000 in the South East and £62,000 in London to a low of £27,000 in Scotland: a ratio of 2.4:1. The differences reflect the regional variation in house price in 1991 and vary from year to year depending on prices, but the overall North–South pattern is stable. Mean equity in the South East is two to three times that in Scotland and about 50% higher than in the Midlands and the North. This finding is not surprising but it confirms that home owners in the South have much greater housing wealth.

There was also a sharp variation in mean estimated equity by grouped SEG of head of household, ranging from £72,000 for professionals and managers to £50,000 for other non-manual workers, £47,000 for skilled manual and unskilled and £42,000 for partly skilled (Table 5.4). These variations parallel the differences in estimated current value of property by socio-economic group. These varied from £103,000 for professionals and managers to £75,000 for other non-manual and £58,000 for the partly skilled. Professionals and managers tend to have higher housing equity than other groups because they have higher incomes and buy more expensive homes, thus enabling them to accumulate more equity over time.

Not surprisingly, there was also a considerable variation according to the age of the head of household, with mean estimated equity rising from £17,000 for 20–29-year-olds to £41,000 for 30–39-year-olds and £67,000 for 40–49-year-olds to a peak of £80,000 for 50–59-year-olds. This figure remains roughly

Table 5.4 Mean equity by SEG (£)

	Adjusted All owners	Estimated		
		All owners	Outright	Mortgaged
Professional	66,500	71,200	122,200	63,400
Managerial	67,700	72,500	134,800	63,900
Other NM	46,100	50,800	86,500	43,300
Skilled manual	42,700	47,400	80,200	39,800
Semi-skilled	32,300	42,400	64,200	34,700
Unskilled	44,600	46,700	63,900	40,100
All SEGs	51,500	55,800	90,600	48,600

Source: British Household Panel Survey (1991)

Table 5.5 Mean equity by age of head of household (£)

Age	Owners' estimated value-based			Adjusted purchase price All owners
	All owners	Outright	Mortgaged	
20–29	16,900	50,600	15,900	14,300
30–39	41,200	84,200	39,400	37,200
40–49	66,600	89,900	62,500	60,100
50–59	80,400	85,200	77,100	83,000
60–69	79,800	84,400	55,900	90,900
70–79	78,500	78,200	83,700	97,600
80+	66,200	66,600	–	82,000

Source: British Household Panel Survey (1991)

constant up to age 79, falling slightly to £66,000 for those aged 80 or over. Mean adjusted purchase price equity is slightly lower for the younger age groups but considerably higher for the 60-year-olds and above as this method produces higher estimates for outright owners and lower estimates for mortgaged owners (Table 5.5).

For mortgagors, estimated equity rises with age from an average of £16,000 for 20–29-year-olds to £77,000 for 50–59-year-olds, then dips to £56,000 for the 60–69 age group and peaks at £83,700 for 70–79-year-olds. The variations reflect a rise in equity with age as the mortgage is paid off and the owner moves to more expensive property. For outright owners the pattern is less marked. Equity rises from £50,600 for 20–29-year-olds to £84,000 for 30–39-year-olds and remains stable at over £80,000 before falling slightly for older outright owners. Age differentials in equity are the most marked of variations, exceeding those of region, socio-economic group or income. They reflect length of time in the housing market and the fact that most older owners have traded upmarket over time and have paid off their mortgages.

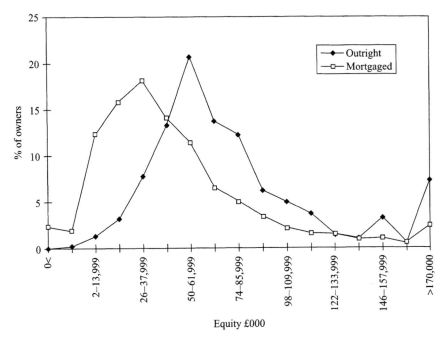

Figure 5.6   Distribution of housing equity of outright and mortgaged owners, 1991
Source: British Household Panel Survey (1991)

Mean Equity among Outright and Mortgaged Owners

The summary of housing wealth outlined above included both outright and mortgaged owners. But these two groups are very different in a number of major respects and they have very different patterns of housing wealth holdings. Outright owners are, on average, considerably older than those buying on a mortgage. The mean age of the head of household for outright owners was 64.4 years compared to 40.9 years for mortgaged owners. Over the age of 55 the great majority of owners own outright (see Hamnett and Seavers (1996) for more detailed analysis).

The difference in the age and outstanding mortgage debt of outright and mortgaged owners is reflected in sharp differences in equity between the two groups. Whereas outright owners have a mean equity of £82,000, mortgaged owners have a mean equity of £49,000. The difference of £33,000 largely reflects outstanding mortgage debt though outright owners do own slightly more expensive properties on average (£82,000) than mortgaged owners (£78,000). The mean outstanding mortgage of mortgaged owners is £27,400. The distribution of housing equity between the two groups is shown in Figure 5.6. Some 28% of outright owners had housing wealth of over £86,000, and 7% had housing wealth of over £170,000. At the other extreme, there is a tiny proportion (4.7%) of outright owners with equity of under £26,000 and 1.4% with equity of under £14,000. The great majority of outright owners have substantial home equity. The distribution for mortgagors is very different. Only 25% had equity over £62,000 and only 6% had equity of over £122,000 compared to 15% of outright owners. Rather surprisingly, there were no differences in housing equity by the gender of the head of household. Male heads accounted for 78.5% of heads and females for 21.5%, but equity was identical at £60,467 for men and £60,829 for women despite the very sharp differences in household income (£18,287 for male-headed households and £7,995 for female-headed households).

Home Owners' Equity in South East England: The MORI Survey

We also used the MORI survey of 972 home owners in South East England to look at the distribution of housing equity in the region. Owners were asked to estimate the sale value of their home and the results show that mean equity for the MORI sample was £79,750 compared to the BHPS figure of £79,875 for the whole country. As house prices in the South East are generally much higher than in the rest of the country this could indicate that BHPS national figures are overestimated or it could simply reflect the fact that the MORI survey was undertaken in late 1993 and early 1994 while the BHPS data relate to 1991. Mean equity in the MORI survey varied by area from a peak of £124,000 in the Chilterns (an area with many large, detached, expensive houses and older owners) to a low of £45,000 in Milton Keynes, an area of smaller, cheaper houses with more first-time buyers. Not surprisingly, equity also varied in line with estimated property value, ranging from £23,250 for houses valued at less than £56,000 to £207,600 for houses worth more than £180,000. Mean equity also varied by socio-economic group, ranging from £99,000 for professionals and £95,700 for managers to £57,300 for the semi-skilled, and varied with year of purchase ranging from £130,000 for houses bought before 1960 to just £25,150 for houses bought in 1989 or after. Finally, it varied according to the number of homes owned, ranging from £61,000 for those who had owned only one home to £123,000 for those who had owned more than three.

The MORI survey found that 5.3% of all owners had negative equity. Overall, it is clear that home equity is generally greatest for professionals and managers, for outright owners, those living in the South East, those in larger, more expensive houses and those who have owned longest, particularly those who bought well before the peak of the 1980s boom. Conversely, equity is generally lowest for those in lower socio-economic groups, those who have bought more recently, those who live in cheaper property, and those who live in the northern regions. The exception to this generalisation concerns those with negative equity who are particularly concentrated in London and the South East. Because of the importance of this phenomenon it is considered in detail below.

The Growth and Importance of Negative Equity in the 1990s

The scale and incidence of negative equity has varied considerably during the 1990s, increasing rapidly in the early 1990s, particularly in the South East, then falling in the mid 1990s as house prices began to rise. It is therefore important to look at the scale and incidence of negative equity both when it was at its peak in the early 1990s and at the more recent pattern. Looking first at the early 1990s, Wriglesworth (1992) and the Bank of England (1992) used similar methodologies using published aggregate data to estimate negative equity. They both involved linking together the extent of regional house price falls since 1988, the number of mortgaged housing transactions, and the average mortgage advance to calculate the likely number of buyers whose house was likely to be worth less than the original mortgage advance.

Using this methodology, Wriglesworth (1992) calculated that in June 1992 there were 1.3 million or 1 in 7 mortgaged households with a mortgage which was greater than the value of their home. The Bank of England estimated that about a million home owners who had bought since 1988 had negative equity in 1992. In London, the South East and East Anglia, where house price falls were greatest, the Bank estimated that two-thirds of the first-time buyers who had bought homes since 1988 had negative equity (Table 5.6). Their estimate was lower because they allowed for the 180,000 properties which had been repossessed by lenders since early 1988. But, as they pointed out, this would be offset by the build-up of mortgage arrears and new advances secured on houses.

Table 5.6   Estimated number of home owners with negative equity in 1992 (000)

	South East	East Anglia	Greater London	South West	Other regions	Total
First-time buyers	260	46	99	87	106	598
Previous owners	121	22	43	49	43	278
Total	381	68	142	136	149	876

Source: Bank of England (1992)

The Bank of England also attempted to estimate the value of negative equity. They calculated aggregate national totals of £3.6 billion for first-time buyers and £2.3 billion for former owners, a total of £6 billion at the end of the second quarter of 1992 making an average of £6,000 per affected household: equivalent to 14% of total personal sector saving in 1991. The shortfall was heavily concentrated in the South East due to the size of the region, but was of proportionately similar magnitude in London, East Anglia and the South West. The Bank noted that nationally 20% of those affected faced notional shortfalls of more than £10,000, but that in the South East some 25% of the affected had shortfalls of over £10,000. In some cases the shortfall would be much greater.

The Bank estimated that if prices were to fall by a further 1% per quarter in all regions, the value of negative equity held by existing borrowers would be £10.5 billion by the end of 1993, with more than 1.6 million households affected. But if prices rose by 1% per quarter, the value of negative equity would fall to £2.5 billion by the end of 1993, leaving about 600,000 households still affected. These were national estimates, however, and they noted that in the South the problem of negative equity was unlikely to be shortlived and suggested that it would require house price inflation of 10% per annum from 1993 to lift all affected households out of negative equity by the end of 1995. They concluded that without rapid price inflation, "negative equity is likely to remain an important feature of the finances of many households for some time to come" (1992: 268).

The estimates from Wriglesworth and the Bank of England were invaluable, but they both had the disadvantage that they were estimated on the basis of regional averages which assumed a uniform fall in price in each region for each household. The true falls have varied around the average, with the result that some households will have more negative equity, offset by others who still have positive equity. The only way around this problem is to examine the situation of individual households and then to aggregate the results. This was attempted by Gentle et al. (1992) using a methodology which grouped households together in order to calculate price falls by house type and market area. They used a sample of almost 1 million mortgage advances made over an 11-year period from 1980 to 1991 by one of Britain's largest building societies. They note that the society, the Nationwide, had "an unrepresentatively higher proportion of mortgages in the South East of England, but lent proportionately less than average in the last few years of the house price boom. Otherwise the sample is thought to be representative of the borrowers of a reasonably typical mortgage lender" (Gentle et al., 1992: 9).

They had comprehensive data on the purchase price of properties and their characteristics, the size of mortgage advance etc., but they had to estimate the current value of properties to calculate the scale of negative equity by using 4 types of housing, 112 housing market areas and 12 years. This enabled them to estimate an average house price for each house type in each area for each year. They assumed their data were representative of all house purchase lending and applied the proportions to national totals. Using this method they estimated that by August 1992, almost 1 million or one in ten mortgage holders had negative equity and almost all (99%) had bought since 1987. The average negative equity was estimated to be £4,400. They estimated that the total value of British negative equity in October 1992 was £2.68 billion and they suggested that, if 1992 price falls were replicated in 1993, the total amount of negative equity would double to £5.2 billion by October 1993.

Gentle et al. found that the geographical distribution of negative equity was highly uneven between North and South. In the South, the proportion of owners who bought between 1988 and 1991 who had negative equity in 1992 varied

116

from 41% (Greater London) to 24% in East Anglia. In the North, the proportion ranged from 8% in the West Midlands and Wales to just 2% in the North and 1% in Scotland. If home owners in the South benefitted disproportionately in the boom years, they also suffered disproportionately during the slump. They also examined which types of area and kinds of housing contain households with highest negative equity. Their results revealed a dramatic "geographical contrast which massively outweighs any traditional social differences between the areas. Those areas which had fared worst were almost exclusively in . . . the poorer parts of the South East and South West whereas those which had suffered least were generally in the poorer parts of Scotland, Wales and the North" (1992: 13).

Within regions, however, the incidence of negative equity was very strongly related to house type, mortgage advance and social class. Thus 27% of buyers who bought homes without central heating (the least expensive category) had negative equity compared to only 4% of those who bought houses with more than three bedrooms. While this might seem surprising, Gentle et al. note that although the more expensive houses have fallen most in price, buyers of these properties generally provided a larger deposit, and have been less affected by negative equity and they found that: "The probability of a household having negative equity is most closely related to the relative size of their deposit. Over half of households with a 5% (or lower) deposit who bought since 1987 now have negative equity compared to less than 1% of those who put down a 20% or higher deposit" (1992: 14).

At a constituency level, the highest proportion of buyers with negative equity were concentrated in predominantly working-class areas in and around northeast and southeast London in areas such as Dagenham, Luton, Basildon, Newham, Woolwich and Leyton where home ownership had recently increased. In some areas the proportion of recent buyers with negative equity exceeded 50%. These were areas in which social housing had declined significantly and in which the Conservative party had recently increased its support.

Dorling (1993) subsequently updated the research to October 1993. His main finding was that, as house prices had continued to fall, so negative equity had increased both absolutely and geographically. Of those who had purchased since 1988 one in four had negative equity by October 1993. The regional figures varied sharply from a high of 40% in London and the Outer South East and 37% in the South West to lows of 5% in Scotland and 6% in Wales. Dorling found that the probability of a household having negative equity was most closely related to the size of mortgage advance. In Britain as a whole, none of those with initital advances of less than 70% had negative equity but this rose to 7% of those with 75–80% advances by the third quarter of 1993. The proportion rose progressively to 74% of those with 100% advances. The average for Britain as a whole was 26%. In southern Britain as a whole the proportions were higher, and in many constituencies in the South East the proportion of recent buyers with negative equity exceeded 50%.

Dorling found that over two-thirds of households with 95% mortgages (or higher) who had bought since 1987 had negative equity. This had risen from 50% in 1992. Younger buyers were most likely to have negative equity as they are most likely to have borrowed highly. Dorling estimated that almost half (48%) of recent buyers aged 20–24 had negative equity in October 1993. The proportion fell steadily to 8% of those aged 50 and over. Average negative equity in London was estimated at £5,300, though this seems relatively small given that it is difficult to purchase anything much under £60,000 in London. Dorling concluded that: "Negative equity affects particular groups of people in certain areas very severely. There is no likely short-term solution to their problems which stem largely from having needed to buy a home at the wrong time in the wrong place." (See also Dorling and Cornford, 1995).

The impact of negative equity has been most marked in and around London, particularly in those areas with a concentration of lower-priced housing. This was clearly illustrated in a report produced for the Association of London Authorities in 1993 which compared average prices in the fourth quarter of 1988 to the average price in the third quarter of 1993 for first-time buyers by borough and found that, because of average price falls of 24%, an average of 79% of first-time buyers who bought in the fourth quarter of 1988 had negative equity. In some boroughs, the proportion was over 90% (Nicholson-Lord, 1993).

More evidence of the incidence of negative equity was provided by a survey of 1,200 home owners in Glasgow, Bristol and Luton undertaken by Ray Forrest and colleagues (1994) in 1993. The choice of areas was designed to provide a mix of areas: Glasgow was a stable housing market, Bristol was moderately volatile and Luton experienced rapid house price inflation in the late 1980s with a sharp fall in the 1990s. The distribution of negative equity paralleled these experiences. In Glasgow only 2.4% of owners surveyed had negative equity compared to 9.5% in Bristol and 25% in Luton. The average for the three towns was 17%. Some 25% of households with negative equity had under £5,000, 44% had between £5,000 and £10,000 and 31% had over £10,000. The age distribution of households with negative equity was strongly skewed, with 46% under 29 years of age and a further 46% between 30 and 39, thus confirming Dorling's findings regarding age.

The incidence of negative equity tended to fall with social class, with 33% of those in class A (professional and managerial) having negative equity, 23% of those in B (technical and other non-manual), 17–18% in C1 and C2 (skilled manual), 14% of Ds (unskilled) and just 8% of those in class E (unemployed, retired). To a large extent, this reflects the fact that those most affected were those who were able to buy in the mid to late 1980s. It should be noted however that, in terms of cases, those in class A only accounted for 6% of the total, and those in classes D and E for 15%. The majority of negative equity was found in classes B (23%) and C1 and C2 (55%). Skilled manual and junior non-manual workers were hit most in numerical terms. A similar pattern was found in terms of income. The incidence of negative equity increased with income, rising to a peak of 25% for household heads earning £400–499 per week.

Equally striking are the findings concerning dwelling type and year of first purchase. Only 12% of those living in detached houses and 15% in semi-detached houses had negative equity, compared to 20% in terraced houses, 37% in purpose-built flats, and 67% of "others", most of which are likely to be converted flats, compared to an overall average of 17%. This suggests that those with negative equity are mostly first-time buyers who bought cheaper property. In terms of council tax banding, 27% of those in Band A (lowest) had negative equity, falling to zero for bands E/F. When negative equity was calculated as a proportion of property value, it ranged from 7% of detached houses to 26% of purpose-built flats and 23% of the "other" category.

The incidence of negative entry by year of entry into home ownership was very revealing. The proportion with negative equity rose from 0% of those who bought in 1985 or before and 5% of those who bought in 1986–87, to 22% who bought in 1988 and 40% of those who bought in 1989–90, falling to 33% for 1991, and just 3% of those who bought in 1992–93. As Forrest et al. (1994) note: "This confirms that date of purchase is a prime determinant of whether a household is in negative equity or not". First-time buyer status was a second influence as first-time buyers are more likely to have taken out loans representing a higher proportion of the purchase price and are thus more exposed to price falls. Overall, 33% of first-time buyers had negative equity but so did 10% of other owners and almost a third of those with negative equity were not first-time buyers.

The Nationwide Building Society (1995) estimated that there were still 1.5 million households in negative equity in 1995, with a total debt of about £7.5 billion. They suggested that the problem remained concentrated in the South of England, which accounted for over half of all cases, and that the average value of negative equity in the South was some £7,000 compared to £2,500 in the rest of the UK. They also noted that as households in the South had been in their homes longer than elsewhere, negative equity may act as a constraint on household mobility. They pointed out there are also an estimated 2.2 million buyers with what they term "neutral equity", that is where equity is not enough to cover the costs of moving house. They suggested their estimates were, if anything, biased downwards as they were based on average house prices whereas the households most affected by negative equity tend to be young, first-time buyers with mortgages of over 90% and lower than average incomes. Price falls for this group have tended to be greater than the average.

The Nationwide estimated that on average 1989 first-time buyers in London had equity of £4,200 in 1989 but negative equity of −£22,300 in 1995. In East Anglia the respective figures are £3,100 and −£18,700. But in Scotland, where the boom was less marked and the subsequent price falls were less sharp, comparable figures were £1,400 and £9,800. By comparison they take the example of a 1979 first-time buyer who traded up in 1986. These show an average equity in London of £1,100 in 1979 and £52,200 in 1986 and £600 and £32,100 in Scotland. First-time buyers in the South East at the top of the boom suffered most severely. As the Nationwide (1995: 5) point out:

> Homeowners' actual position depends crucially on the timing of trans-
> actions and local market conditions: negative equity has been unavoid-
> able for those unlucky enough to buy at the wrong time or in the wrong
> place. The majority of homeowners with negative equity have simply been
> unlucky in that the timing of their first house purchase was near the top of
> an inflationary boom, shortly before the bubble burst.

More positively, the Nationwide suggested that relatively small price rises of 2–3% would remove 50% of home owners from negative equity in Yorkshire and Humberside, the North, the West Midlands and Wales, and that rises of 7–8% would remove 90% of owners from negative equity in these regions. On the other hand, price rises of 10% would be needed to remove 50% of owners from negative equity in London and the South East, and price rises of around 20% to remove 90% of owners in these regions. Fortunately, price rises in London and the South East have been substantial in 1996–7 and a large proportion of owners in these areas have been removed from negative equity. The scars will remain, however, and are unlikely to allow a rapid change of attitudes towards home ownership and accumulation. Although a great majority of home owners still have considerable positive equity in their homes, they are the fortunate ones who bought early or on lower mortgage advance to price ratio. A generation who bought at or near the peak of the boom in southern England have been badly burnt, and several hundred thousand have gone into mortgage arrears or suffered repossession when they were unable to meet their mortgage payments. As earlier analyses showed, housing equity is strongly related to age, social class, region, date of entry into the housing market, property type and price. Not surprisingly, housing equity is generally highest for managers and professionals, those who own expensive houses, particularly in southern Britain, and those who have owned for longest, paid off most of their mortgage and had most of the years of capital accumulation in the 1970s and 1980s. Equity was lowest for younger households, those in the northern regions and those who bought at or near the peak of the 1980s boom and saw house prices fall. As with capital gains, timing in the housing market is very important.

Kaletsky (1995) disagrees with these analyses, arguing that the importance of negative equity has been considerably overstated and that people can and do escape from negative equity even if house prices fall by the simple expedient of paying off their mortgage "through the long-steady drip-drip of mortgage repayments or endowments". Kaletsky is correct in this, but seems to neglect the short-term problems caused for people who are unable to continue their repayments by virtue of unemployment or need to sell their property if they split up. Paying down the mortgage is fine if you have a job and the income to do it.

Many did not, hence the wave of repossessions. Negative equity also affects housing mobility in that, when you sell, negative equity is crystallised into real losses. Along with repossessions, negative equity was one of the great problems of the home ownership market in the early 1990s: a problem from which many owners have still not recovered.

Negative equity has devastated many new home buyers. Rather than being a route to long-term capital accumulation, as Saunders envisaged in the late 1970s, home ownership has proved to be a financial albatross around their necks. It is still true that for most owners who bought their homes after the War, and certainly from the 1960s onwards, ownership has proved a source of long-term capital gains. But for a generation of buyers in southern England who had the bad luck to buy a home "at the wrong time in the wrong place" as Dorling tellingly puts it, the effect has been severe.

CHAPTER 6

Housing Inheritance and Equity Extraction from Home Ownership

We have seen in Chapters 4 and 5 that home ownership has been a source of considerable capital gains over the past twenty to thirty years for those owners who did not enter the housing market or move upmarket at the peak in the late 1980s. House prices are always expensive relative to income, but while incomes rise, outstanding mortgage debt generally decreases over time. Eventually, unless household breakup, repossession or some other disaster intervenes, the owner pays off the mortgage and becomes an outright owner. There are currently 6 million outright owners in Britain today, most of them in their fifties or older, as Chapter 5 showed. Mortgaged owners also accumulate equity. But what happens to all this accumulated equity? How and when is it released and who benefits? This is the subject of this chapter.

Until the early 1980s most outright owners were unable to do very much with their asset except live in it and eventually bequeath it to beneficiaries, whether to their children or other relatives, charity or the local dogs' home. As for those paying off the mortgage, the main objective was simply to make the payments on time. While it was possible to extract some equity by moving downmarket or by taking a bigger mortgage than strictly necessary on moving, large-scale equity extraction was uncommon, not least because the building societies were unwilling or unable to lend money against the property to be used for purposes other than house purchase or improvement (and detailed bills were required). Most housing equity remained intact until death and eventual inheritance. But, since the liberalisation of lending associated with financial deregulation in the early 1980s, there has been a rapid expansion of equity extraction by existing owners either in the process of moving or by simply borrowing against security of the property. As Chapter 8 shows, this proved a considerable problem for the management of the economy in the late 1980s as the boom in housing equity release helped to fuel consumer spending by asset-rich owners. Later in this chapter I discuss the scale and extent of equity release and its implications for housing inheritance but first I look at housing inheritance directly.

Housing Inheritance in Britain

House inheritance has been a fact of life for generations among the better off middle and upper classes. The importance of housing inheritance is clearly

123

shown in novels ranging from Trollope's Pallisers to Galsworthy's *Forsyte Saga*, which includes the aptly titled *The Man of Property*. The "reading of the will" was a major event in the nineteeth century among the middle and upper middle classes as it determined who got what. The working classes did not figure much in the novels of inheritance, largely because they had little to inherit. Inheritance, like wealth, was concentrated in the hands of a small minority of the relatively well off (Harbury and Hitchens, 1979). Most wealth was transmitted intergenerationally rather than built up by the fruits of own's own labour or that of others (Harbury, 1962; Harbury and Hitchens, 1976).

The growth of home ownership and the rapid rise in house prices during the 1970s and 1980s prompted a sudden realisation by many observers that the expansion of housing wealth was likely to cascade down in an expansion of housing inheritance as the new postwar generation of home buyers died and bequeathed their assets to their children. This was reinforced by the finding that an increasing proportion of the elderly were now home owners. In 1951 there were 5.5 million people aged 65 and over (11% of the population). By 1981 the number had risen to 8.5 million (15.5%) and by 1991 it had risen to 9.9 million. The size of the elderly population is projected to remain static in the 1990s but the rates of home ownership among the 65-plus age group have risen from 44% in 1975 to 54% in 1990, and should rise to over 60% by 2000 as those in middle age (who have higher rates of ownership) move into the older age groups. The views of the proponents of the "mass inheritance" thesis in the late 1980s are illuminating. Peter Saunders (1986: 158) stated that:

> With 60% of households in the owner occupied sector in Britain and an even higher proportion in other countries such as Australia and the USA, not only is a majority of the population now in a position to accumulate such capital gains as accrue through the housing market, but for the first time in human history we are approaching the point where millions of working people stand at some point in their lives to inherit capital sums far in excess of anything which they could hope to save through earnings from employment.

A report on housing inheritance produced by Morgan Grenfell (1987) argued:

> The effect of increased property ownership in the 25–40 age group during the fifties and sixties is now being reflected in high owner occupation rates for the retired population. Half of heads of households in the UK over 65 years of age are now owner-occupiers; this is likely to approach two-thirds by 2000. As a result there will be a large increase in property inheritance accruing to a majority of households over the coming years.

And in an editorial, "Growing Rich Again", *The Economist* (1988a) argued that:

> Britain's middle classes . . . are about to grow rich once more. Two mechanisms are changing the way money is distributed in Britain. One is inheritance . . . The generation of Britons now reaching retirement age was the first to

put a big proportion of its savings into home ownership. Today 50% of pensioner households are home owners; by the year 2000 that figure may be 60% and still increasing. These pensioners, often caricatured as poor, are really growing rich, but their riches are tied up in the roofs over their heads. Only when they die, bequeathing their property to their children, can this most popular form of British investment be cashed in. So the bequest, that staple of Victorian melodrama, is about to make a come-back.

Finally in 1988, Nigel Lawson, then Chancellor of the Exchequer, stated that "Britain is about to become a Nation of Inheritors. Inheritance, which used to be the preserve of the few, will become a fact of life for the many. People will be inheriting houses and possibly also stocks and shares."

It is clear from these statements that most commentators saw Britain becoming a nation of inheritors, where, to quote John Major, wealth "cascades down the generations". Britain was seen to be on the brink of a major social revolution which would lead to a wave of housing inheritance affecting a large proportion of society, particularly the middle classes whose parents had the foresight to buy good homes in the interwar years or in the 1950s and 1960s. The purpose of this chapter is to examine the evidence on housing inheritance, its scale and value, in an attempt to shed light on the validity of these assertions. It will be argued that while inheritance is an important phenomenon, most commentators in the 1980s fell into a trap of assuming that home owners leave their property to their beneficiaries and would not seek to extract any of the equity from the property prior to death. But extraction of housing equity has risen considerably in importance during the 1980s and some owners now have to sell their home to pay for private residential care. The importance of equity extraction lies in the fact that the amount of equity tied up in housing at any point in time is finite. It can either be transmitted via inheritance or be extracted prior to death, but not both. Put simply, the greater the level of equity extraction, the less there is available for inheritance. The more of one, the less of the other.

There are two ways of assessing the scale and value of housing inheritance. The indirect method is to make a series of calculations based on the incidence of home ownership by age, death rates by age, the proportion of married home owners and average house prices, corrected for age. This method is necessary for making forward projections. The direct method is to analyse Inland Revenue statistics on the number and value of estates passing at death which include house property. Both methods have their problems. The first is reliant on the accuracy and comprehensiveness of the assumptions, while the Inland Revenue statistics exclude house property passing between spouses, property which is transferred prior to death or property placed in trust in order to avoid tax.

One of the first projections of the scale of house inheritance, by Morgan Grenfell (1987), estimated a total of 155,000 estates including house property in 1986 rising to 202,000 by 2000. A second, more detailed projection (Hamnett, Harmer and Williams, 1991), based on population projections by age group,

Table 6.1 Projections of the scale of housing inheritance: non-spouse transfers (£000 p.a.)

	1986	1990	1995	2000	2005	2010
Morgan Grenfell (1987)	155	160	178	202	–	–
Hamnett et al. (1991)	–	168	188	207	227	246
Morgan Grenfell (1993)	111	128	180	200	220	235
Lloyds Bank (1993)	–	133	139	153	–	–
Westaway (1993)	153	106	167	182	201	221

Source: Author's own

home ownership rates and death rates, arrived at similar figures, projecting 168,000 cases a year in 1986–91, rising to 207,000 a year in 1996–2001 (Table 6.1). These projections have proved very optimistic to date for reasons which will be explained below.

Measuring Housing Inheritance: The Inland Revenue Statistics

The basic source of information on the scale, value and asset composition of estates passing at death is produced annually by the Inland Revenue. Since 1968–69 the IRS have given figures on UK residential property as a separate category. This information is collected for the purposes of taxation rather than analysis of inheritance, and the figures are based on a stratified sample of applications by executors for grants of representation or probate for deceased persons' estates. This source has advantages and disadvantages. On the positive side, it is consistent and comprehensive in terms of the categories it collects data for. As estates over a certain changeable minimum value require a grant of representation, a grant of probate or letters of administration before they can be legally administered or distributed to beneficiaries, evasion is limited and the figures are comprehensive. No probate, no inheritance. The value of estates may be depressed by the transfer of assets prior to death to avoid inheritance tax, but this is more likely to result in a reduction in the value of assets rather than failure to enter the statistics at all. The problems with the Inland Revenue Statistics concern exclusions, not avoidance. There are two main exclusions from the IRS statistics. First, joint property passing between spouses is exempt from tax and a formal account is not always submitted for exempt property. Second, orders made under the Administration of Estates (Small Payments) Act 1965 permit small estates containing certain assets up to a value of £1,500 (raised in 1984 to £5,000) to be dealt with without production of a grant. As many people, such as tenants, have little in the way of assets, these two exclusions together account for the fact that the Inland Revenue statistics on "estates passing at death" average around 270,000 per annum or only 40% of the 660,000 deaths a year.

This may seem a significant omission, but as the excluded estates are either small estates with assets under £5,000 or joint property passing between spouses, the exclusions are unlikely to have a significant effect on housing inheritance as they do not involve a significant intergenerational transfer of assets. As the IRS note: "An excluded estate can include a dwelling only if it is owned jointly in such a way that the deceased's share passes automatically to the surviving joint owner" (1986: 100). Because it is now more common for property, particularly house property, to be owned jointly by a husband and wife rather than just by the husband, it is likely that there are now more excluded estates involving house property than there were twenty or thirty years ago. This will delay, but not eliminate, entry of property into recorded statistics as surviving joint owners eventually die.

The prophets of an "inheritance economy" suggest that the scale of housing inheritance is likely to grow rapidly in the future. And, as home ownership and the proportion of older home owners have been growing rapidly since the 1950s, there is a strong implication that the scale of housing inheritance should have increased since the 1970s. But Inland Revenue statistics show no evidence for a long-term increase in the number of "estates passing at death" including housing. On the contrary, the number of estates containing house property has been broadly stable. In 1968–69, the first year for which IR data on housing in estates became available, 271,200 estates were recorded. Of these, 125,000 or 46% contained house property. The number of estates containing house property rose to 149,600 in 1969–70 and has remained broadly constant at this level every year since although it rose to 160,500 in 1993–94 (Table 6.2). The pattern can be best described as one of "trendless fluctuation" and it is clear that there has been no increase in the scale of housing inheritance over the twenty-five years. The number of recorded estates has been stable at around 270,000 per annum and the number including house property has been stable at around 145,00 per annum, although the proportion of estates containing house property has risen from 46% in 1968–69 to 56% in 1993–94. All the figures given above are for the number of estates passing at death. But it should be stressed that not all these estates are available for inheritance. A substantial number of them pass to surviving spouses even though the property is not jointly owned. The scale of these transfers can be estimated from the Inland Revenue data on the marital status of the deceased. About 48% of male estates and 16% of female estates belong to married persons: an average of 39% of all estates. While some property may pass to children or other beneficiaries, the bulk of the estate and the family house is likely to pass to the surviving spouse. This reduced the number of estates with house property available for non-spouse transfers to approximately 87,000 in 1992–93 (Holmans and Frostega, 1994). This figure is far below the projections made by the proponents of mass housing inheritance.

This does not apply to the value of house property, of course. Measured in current prices, the value of housing in estates rose steadily from £465 million in 1968–69 to a peak of £9.46 billion in 1989–90, since when it has decreased, as

Table 6.2 The number and value of estates passing at death

	Number of estates with UK dwellings	Total number of estates	% of total with dwellings	Value of housing (£m)	Total value (£m)	Housing as % of total
1968–69	125,085	271,238	46.1	465	1,923	24.2
1969–70	149,592	287,239	52.1	501	1,948	25.7
1970–71	142,473	267,718	53.2	530	1,967	26.9
1971–72	149,052	288,796	51.6	638	2,275	28.0
1982–83	143,980	288,199	50.0	3,383	8,211	41.2
1983–84	148,800	296,890	50.1	3,683	9,195	40.0
1984–85	147,717	273,762	53.9	4,163	10,372	40.1
1985–86	137,486	245,071	56.1	4,567	11,482	39.8
1986–87	149,265	270,947	55.1	5,398	12,738	42.2
1987–88	125,532	234,688	53.5	6,020	14,310	42.1
1988–89	140,561	249,233	56.4	8,439	17,320	48.7
1989–90	154,225	276,412	55.8	9,460	20,121	47.0
1990–91	140,561	249,233	55.5	8,579	18,580	46.2
1991–92	143,494	255,172	56.2	8,347	19,453	42.9
1992–93	142,446	254,450	56.0	8,016	19,511	41.1
1993–94	160,512	285,125	56.3	8,853	22,196	39.9
1994–95	150,807	270,868	55.7	8,567	21,758	39.4

Source: Inland Revenue Statistics (various years)

a result of the slump in house prices in the early 1990s, to £8.85 billion in 1993–94. In percentage terms, the value of housing in estates rose from 24% in 1968–69 to a peak of 49% in 1988–89, before falling back to 40% in 1993–94. As the number of estates has not increased, the rise (and fall) in the value of house property in inheritance is entirely a result of house price inflation and recent price falls. The value of housing inheritance has increased almost 2,000% in just over twenty years. The value of inheritance overall also rose from £1.92 billion to £20.12 billion: an increase of 950%. Non-housing inheritance increased in value by only 632%. The value of housing inheritance therefore rose twice as fast as inheritance in general and by 3.2 times the value of non-housing assets. The increase in the total value of inheritance over those 20 years owes a great deal to the increase in the value of housing assets. As housing makes up over 40% of the total value of estates, it has been the principal driving force behind the increase in the value of estates. Notwithstanding the fall in the value of house property in estates in the early 1990s, it still accounted for 40% of the total value of estates in 1993–94 greater than any other asset.

The distribution of estates by size containing house property differs from the distribution of all estates. Although few estates worth under £25,000 contain house property, house property is then strongly represented in the middle-value estates from £25,000 to £100,000. The proportion of estates with house property is lower than the proportion of all estates in the larger size bands. The

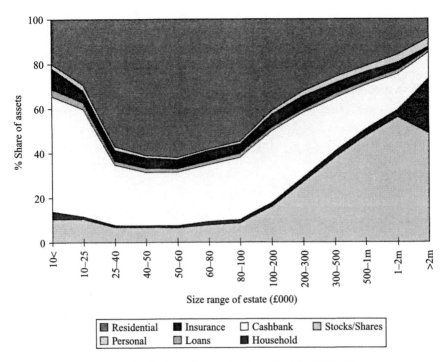

Figure 6.1 Percentage distribution of assets by size of estate, 1993–94

Source: Inland Revenue Statistics (1996)

distribution by value of all estates, and house property is very similar, however. Figure 6.1 shows that the number of estates containing house property is greatest in the small to medium wealth bands from £10,000 to £200,000. These bands contain 91% of estates including house property. The concentration by value is slightly less (81%). The concentration by value is at its peak in estates valued at £60,000–200,000. But the distribution of house property in estates is generally more evenly distributed than other assets although there are, of course, wide variations. The average value of house property in estates in 1993–94 was £55,150 against £61,000 in 1989–90. These figures are averages, however, and there are large variations in the value of house property between estates. In estates of £25,000–40,000, the mean value of house property in 1993–94 is £30,800, rising to £72,000 in estates worth £100,000–200,000 and to £537,000 in estates worth over £2 million. As ever, the few inherit far more than the many.

A Nation of Inheritors? Who Inherits House Property

All the figures given above relate to the number and value of estates passing at death. But the number of beneficiaries is far higher. A survey of beneficiaries

Table 6.3 The incidence of household housing inheritance by SEG and tenure, 1991

	A	B	C1	C2	D	E	All
% of households who have inherited housing	16.4	16.9	10.9	9.7	4.8	4.1	9.1
	Mortgaged	Outright	All owners	Council tenants		Private	Total
% of households who have inherited housing	7.5	21.7	12.7	2.1		3.9	9.1

Source: Hamnett (1991)

carried out by NOP for Hamnett and Williams (1993) found that the number of major beneficiaries per estate averaged three. This suggests that the mean value of house property inherited is more than £20,000 per beneficiary and, whereas those inheriting house property in the lowest value estates may inherit perhaps only £10,000 each, those inheriting more valuable property may inherit sums of £100,000 or more. Given that the number of beneficiaries varies from one to four or five the actual sums will vary considerably.

Peter Saunders (1986) suggested that "millions of ordinary working people" will inherit property. Other commentators have been less sanguine. Both Morgan Grenfell (1987) and *The Economist* (1988a) suggested that housing inheritance would tend to accrue to existing home owners, particularly to middle-class owners. Evidence from two large national surveys of beneficiaries carried out in 1989 and 1991 (Hamnett, 1991; Hamnett et al., 1991; Hamnett and Williams, 1993) show that the great majority (around 80%) of current beneficiaries are existing home owners. They also show that, while the majority of beneficiaries are in the (numerically larger) intermediate and junior non-manual and skilled manual occupational groups, the incidence (or probability) of inheritance is far higher among the professional and managerial socio-economic groups than amongst other groups (Table 6.3). While 9% of all households interviewed had inherited a share of house property or proceeds, the proportion for professionals and managers was 16.8% falling to 15% of other non-manual households, 11% of skilled manual households and just 4% of the semi-skilled and unskilled. The incidence of inheritance is also higher in the South East (16%), the South West (11%), East Anglia (11%) and Wales (10%) than it is in other areas of the country with Scotland (5%), London (8%) and Yorkshire and Humberside (6%) having rates of housing inheritance half the former. Why should this be so? The explanation is simple and is a result of the parental class and tenure backgrounds of beneficiaries rather than beneficiary characteristics. The higher incidence of housing inheritance in southern England reflects much higher rates of ownership in the south in the interwar period onwards, while London was dominantly privately rented and Scotland had high rates of private and council renting (Hamnett, 1992a; Harmer and Hamnett, 1992).

The current distribution of housing inheritance reflects the characteristics of parental home owners a generation ago when home ownership was much more restricted to the middle classes than it is today. Because the children of middle-class home owners are themselves more likely to be middle-class home owners, the current distribution of housing inheritance reflects the class distribution of ownership thirty or forty years ago (Hamnett, 1991). It can be argued as a corollary to this that the distribution of housing inheritance will widen considerably over the course of the next twenty to thirty years as the current generation of home owners leave property to their beneficiaries. There will, however, still be one group who are much less likely to inherit: the children of today's council or private tenants. And, because the children of council tenants tend to marry children of tenants, they are less likely to share in any spouse inheritance. Thus, although perhaps 70% of individuals today stand to inherit a share of house property or proceeds at some stage in the future (assuming such property is still available for inheritance), there are likely to be 30% of households who will not inherit. Watt (1993), Hamnett (1991) and Thorns (1994) discuss the class basis of inheritance in more detail.

Explaining the Shortfall in the Scale of Housing Inheritance

The Inland Revenue figures present us with a paradox. On the one hand the growth of the elderly population and the rising number of older home owners point towards an increase in the number of estates including house property. Projections of the likely number of estates containing house property suggested a figure of around 180,000 in 1995 (Morgan Grenfell, 1987, 1993; Hamnett et al., 1991). Lloyds Bank (James, 1993) came in with a lower figure of 139,000 and Westaway (1993) with a figure of 167,000. But the IRS data provide no evidence for this. How then can the discrepancy be explained? There are three possible explanations.

First, it could be argued that the great expansion of home ownership since the War has not yet worked through into the dying population. This is most unlikely. Projections of the future of housing inheritance explicitly take into account the changing age structure of the population and age-specific death rates, and the growth of elderly home owners is slowly working through into deaths. The second possibility is that a growing number of elderly home owners own houses jointly with spouses, and that the property is passing automatically to the surviving spouse, only entering the statistics on the death of the spouse. This is very likely, but its effect will only be to delay the passing of the estate until the death of the spouse, rather than reducing the number of estates.

The third and most likely possibility is that some owners are selling or transferring ownership of their houses prior to death, either to pay for residential care, to provide income in old age, or to avoid paying for care. The flaw with almost all comments made in the 1980s regarding the growth of inheritance was that they all rested on the implicit assumption that, when elderly owners died, they died with their property intact and available for inheritance. The possibility that many elderly home owners might dispose, voluntarily or otherwise, of their property prior to death was not really appreciated.

This was a classic case of perceptions failing to keep pace with reality. As is now well known, the 1980s saw a massive increase in private residential care in Britain. Between 1980 and 1991 the number of private residential homes in Britain increased more than threefold from 2,000 to over 9,000 and the number of private residential and nursing home places doubled from 140,000 in 1986 to 286,000 in 1991. Initially, the government picked up a substantial slice of the costs via income support payments, and expediture on this head rose from £10 million in 1979 to £459 million in 1986 and £1,870 million in 1991. Largely to rein back expenditure, the Community Care Act was introduced in 1993. Amongst other things, this replaced direct government support via income support payments with local authority assessment of individual need, and subsequent payment for care where deemed appropriate and necessary (Hamnett, 1995b and 1995c).

The current situation is that individuals with capital assets of over £16,000 are not eligible for government financial assistance for care, those with assets of £8,000–16,000 can receive assistance on a sliding scale, and those with assets of less than £8,000 are eligible for full assistance. Similar rules apply for help with the cost of local authority or voluntary sector homes: those with assets over the threshold level must pay a sliding contribution. The implications of these developments are very profound, particularly when combined with the rapid closure of many NHS long stay hospitals, geriatric and other units in the last few years. Increasingly, elderly people in need of long-term care must pay for it if they have capital assets. While it is possible to pay for the costs of care out of income or savings, only a small minority of the elderly are able to do this, as care costs average between £13,000 and £25,000 per annum. The other options are for family or friends to pay, or for capital assets to be run down or sold to meet the bills.

While it is possible for the more affluent to sell stocks and shares or realise other assets, for most people their house is their major asset and calculations suggest that a growing number of homes are being sold each year to pay for care. While it is possible to estimate the number of houses sold each year to pay for care, it is impossible to estimate the number of houses transferred to children prior to death or sold while the owner is in good health to avoid payment for care. It is not something people are generally willing to discuss, and no figures exist. Anecdotal evidence indicates, however, that some elderly people or their prospective beneficiaries are aware of the rules about assets and payment for care and seek, where possible, to avoid payment by early transfer of property and other assets.

To estimate the number of homes that may be sold for care it is necessary first to gauge the size of the residential and nursing home care sector. In the early 1990s, the total number of private residential and nursing home places in

Britain was approximately 300,000. The number of voluntary places was 54,000 and there were 120,000 local authority places: a total of 470,000 places. Laing and Buisson (1991) have estimated that the average length of stay in care homes is 2.5 years, which suggests that the average turnover of residents is about 40% a year. Some, of course, will stay for much longer and others for a much shorter time, but this indicates 190,000 new entrants into care homes each year assuming full occupancy. A 90% occupancy rate would suggest 170,000 new entrants. Even a conservative estimate of 25% turnover would indicate 120,000 new entrants a year. Given that about 50% of people aged 60 and over and living alone now own their own homes, we can estimate that the number of home owners entering care homes each year is between 60,000 and 85,000. This is likely to be reduced to an unknown extent by transfers of house property to children and other potential beneficiaries. On the assumption that two-thirds of home owners entering care have to sell their homes to finance care, this suggests that 40,000–56,000 elderly owners may sell their homes each year to pay for care. If only half of home owners have to sell, the figure would still be 30,000–43,000 a year (Hamnett, 1997b).

Holmans (1997; see also Holmans and Frostega, 1994) has estimated that 30,000 older owner occupied households were dissolved in 1990 through moves to live with relatives or into institutions. This should be taken as the lower bound, and in cash terms it would represent an equity withdrawal of about £1.8 billion a year assuming that the homes are sold. These estimates represent a substantial reduction in the number of homes available for inheritance and help explain the discrepancy between the projected levels of housing inheritance and Inland Revenue figures. The shortfall between projected numbers of houses passing at death and the Inland Revenue figures on the number of estates passing at death containing house property is currently about 36,000–46,000 depending on which projection is taken. This is in line with the estimates of the number of homes being sold in order to pay for care or transferred prior to death. This helps explain why the projected boom in housing inheritance has not come to pass. Britain is unlikely to become a nation of inheritors, now or in the near future, although Westaway (1993) anticipates a sharp increase in the scale of inheritance from the mid 1990s onwards as the post-World War One baby boomers die.

Equity Extraction

Most home owners who buy with a mortgage go through a process of equity accumulation and eventual equity release. As more and more of the mortgage is paid off, so they have greater equity in the home until the day comes when owners are "Dun Payin". Until very recently, the equity remained locked up until the owner died and the house passed to their beneficiaries. Houses functioned as a store of wealth over the life cycle and across the generations. The amount of equity released each year by so-called "last-time sellers" plus any accumulated equity taken out in the process of moving house (to pay for moving costs, new carpets, fridges etc.) should equal the amount of equity put in by new buyers as deposits plus mortgage lending for house purchase and improvement. The only way of extracting equity from a house during life was to move down-market to a cheaper property and keep the difference (Jones, 1978), or sell up and move abroad or move into rented housing or in with family or friends.

The idea that home owners could take out larger mortgages than they needed and use the surplus for consumption was more or less unheard of and building societies simply would not generally lend for this sort of thing. Mortgages were often rationed and only available to those with accounts of more than one or two years' standing, and mortgage waiting lists were common. It therefore came as something of a suprise when equity extraction or "capital leakage" was first identified as an issue in the late 1970s, with owners taking out larger mortgages than necessary in order to finance purchase of consumer goods. Kilroy (1979), Leigh Pemberton (1979) and Downs (1980) argued that excessive lending for house purchase and equity release was leading to increased consumer spending.

The possibility of large-scale equity extraction was limited in Britain until 1979 because of a formal memorandum of agreement between government and the building societies which effectively restricted them to lending only for house purchase or bona fide improvements. But when the Conservatives were elected in 1979 they allowed the agreement to lapse, and in 1980 the "corset" which had restricted the development of bank lending and borrowing was lifted; then the banks entered the mortgage market and the 1986 Building Act allowed societies to obtain access to wholesale funding and effectively ended mortgage rationing. The key elements were in place for the liberalisation of lending and, in retro-spect, for a boom in equity extraction as some owners decided to increase their consumption by reducing the value of their assets or borrowing against them (which amounts to the same thing). Some economists suggest that many home owners had previously been forced to hold more of their assets in the form of housing than they wanted to and that liberalisation of mortgage lending per-mitted them to adjust their asset portfolio (Miles, 1992a).

The first signs of official concern in Britain occurred in 1982. Davis and Saville (1982: 396) pointed out in the *Bank of England Quarterly Review* that: "Concern about the possibility of direct withdrawal of equity from housing by borrowers obtaining more finance than required for house purchase, and its possible implications for credit and monetary aggregates, prompted a request to mortgage lenders by the Bank of England and the Treasury in January 1982 to limit this possibility."

An article in the *Bank of England Bulletin* by Drayson (1985) noted the very rapid growth of what was termed "net cash withdrawal" (NCW) from the private housing market. This was defined as net lending for house purchase that was not applied to either new private sector housing or improvement of existing pri-vately owned homes. The article suggested that the volume of NCW had grown rapidly from £1.54 billion in 1979 to £5.7 billion in 1983 and £7.2 billion in

1984, when it accounted for 43% of net new loans for house purchase, and many commentators interpreted the figures as a direct measure of the money taken out of the housing market for general consumption. Martin Pawley (1985) argued that equity extraction was now the main motor of the British economy. We have, he suggested:

> found a way not only to get rich from our own houses, but perhaps even to live off them entirely in the future . . . The nation of home owners has begun to play Monopoly in Reverse . . . Unlike the traditional board game, in which players start off with property and try to turn it into cash . . . The new style Monopoly player will begin to do consciously what he or she has done instinctively ever since they got a foot on the "ladder" of home ownership. That is to increase the share of annual expenditure that is drawn from long-term housing credit, at the expense of the share drawn from earnings.

The argument is a fascinating one, but unfortunately for home owners, home ownership is not a perpetual financial energy machine that produces money from nothing (despite a feeling to the contrary in some quarters in the 1980s). This is no more than a contemporary version of the old alchemists' dream of turning base metals into gold, and it rests on a fundamental misunderstanding of the nature and origins of net cash withdrawal.

The owner occupied housing market consists of a stock of dwellings, an inflow of new households into the sector and an outflow of households either by death, household amalgamations, movement into other tenures or migration abroad. Whereas new households have to inject equity via a deposit and by borrowing using a mortgage, households leaving owner occupation generally take some equity with them (or leave it to their beneficiaries in the case of death). Housing equity extraction is therefore a normal and inevitable part of the home ownership cycle. Households inject equity when they first enter, accumulate equity during their housing career as they pay off their mortgage, and extract equity when they trade down or exit the sector altogether. As the home ownership market grows and as house prices rise, the scale of equity extraction will also rise, as will the amount of mortgage finance needed to underpin the market. As Davis and Saville (1982: 395) comment, equity extraction from the housing market "is inevitable; every chain in the secondhand housing market has an end; the final house comes onto the market because its owner occupier has died, or ceased to own his [*sic*] own house for other reasons, or because it is put on the market by its landlord after the tenant has left".

Each chain in the housing market also has a beginning, and equity extracted from the secondhand market is balanced by the deposits from first-time buyers (which became of much less importance in the late 1980s), the heavy mortgage repayments made by buyers in the early years of purchase and by injections of mortgage finance by the institutions. For each seller there is a buyer, and every pound extracted from the housing market is paid for directly or indirectly by a

Table 6.4 Housing transactions with no intermediate equity withdrawal (£)

	Receives from sale	Pays off mortgage	Borrows new mortgage	Purchase price	Equity contribution or withdrawal (–)
1st time	–	–	25,000	30,000	5,000
2nd buyer	30,000	–15,000	35,000	55,000	5,000
3rd buyer	55,000	–25,000	60,000	90,000	–
last time	90,000	–20,000	–	–	–70,000
Totals		–60,000	120,000		–60,000

Source: Lee and Robinson (1990) table 1A

buyer. There is no free lunch in owner occupation. Instead there are inter-generational transfers of wealth from young to old and from tax-paying non-owners to mortgaged owners. All equity outflow is matched by an equivalent equity inflow. The capital gains made by last-time equity extractors are paid for by a generation of new buyers, paying a high proportion of incomes.

This process of equity injection and withdrawal into the housing market has been clearly shown by Lee and Robinson (1990) in the context of "chains" of transactions which link buyers and sellers. Table 6.4 shows an imaginary chain of housing purchases and sales with four links. A first-time buyer buys a house for £30,000, using a mortgage of £25,000 and a £5,000 deposit. The vendor uses part of the money to pay off the old mortgage of £15,000, takes out a new mortgage of £35,000 and buys a house for £55,000, injecting another £5,000 from savings towards the price. The new vendor also increases her mortgage substantially by paying off the old £25,000 mortgage and taking out a new £60,000 mortgage towards the purchase price of £90,000. In the example, this last property is sold by the trustees of the owner, who pay off the outstanding mortgage of £20,000 and put the balance of £70,000 into the estate. Lee and Robinson note that the net effect of these transactions is that new borrowing of £120,000 has financed repayment of £60,000 of old loans and equity withdrawal of £60,000.

Table 6.4 focused on the last-time seller. Lee and Robinson's next example (Table 6.5) illustrates equity withdrawal by moving owner occupiers. They assume that five years have passed since the transactions in Table 6.4 took place, that all prices are 50% higher, that building societies and banks are willing to lend on more generous income multiples and that each seller now has a nominal profit on the sale, of which most but not all is put back into the housing market. The fourth transactor in the table is no longer a last-time seller but someone who trades down and launches a further chain which leads to a large increase in borrowing net of repayments which finances substantial equity withdrawal.

Lee and Robinson note that in their second example everyone behaves in a "responsible" way, in that the withdrawal of equity by each transactor is less than their profit on the earlier transaction and the loans are covered by housing

Table 6.5   Housing transactions with intermediate equity withdrawal (£)

	Receives from sale	Of which "profit"	Pays off mortgage	Borrows mortgage	Pays for purchase	Equity contribution or withdrawal (−)
1st time	−	−	−	40,000	45,000	5,000
2nd time	45,000	15,000	−25,000	65,000	82,500	−2,500
3rd time	82,500	32,500	−35,000	95,000	135,000	−7,500
4th time	135,000	52,500	−60,000	−	45,000	−30,000
5th time	45,000	15,000	−25,000	65,000	82,500	−2,500
6th time	82,500	32,500	−35,000	95,000	135,000	−7,500
last time	135,000	−	−60,000	−	−	−75,000
Totals			−240,000	360,000		−120,000

Source: Lee and Robinson (1990)

collateral. Nonetheless, the result is "a substantial injection of extra money into the economy at each link in the housing chain". They also note that the number of opportunities for cash withdrawal is linked to number of sales in the housing market. The greater the turnover, the greater the equity extraction potential. In a period when more houses are being sold, as in the late 1980s, the potential for intermediate equity extraction is enhanced.

Measuring the Scale of Equity Extraction

Various attempts have been made to measure equity extraction (Kemeny and Thomas, 1984; Lowe, 1989; Davis and Saville, 1982; Holmans, 1986, 1991; Miles, 1992a; Westaway, 1993). I use Holmans's (1991) figures as they are the most comprehensive source. Holmans argues that housing equity withdrawal is complex and comprises several different kinds of financial flows, generated in different ways and with different effects. He states that, while it is relatively straightforward to estimate net equity withdrawal by juxtaposing net lending for house purchase with capital expenditure on owner occupied housing, it is crucial to distinguish gross flows if answers are to be given to questions as: "how much of the increase in total housing equity withdrawal was due to withdrawal by last-time sellers, how much by moving owner occupiers over-mortgaging, and how much by owner occupiers increasing their indebtedness secured on their dwellings by adding to their existing loans or remortgaging" (1991: 7).

Holmans distinguishes four separate categories of withdrawal. These are: (a) last-time sales where the seller does not use the proceeds to buy another house, (b) trading down by moving owners to a smaller house or cheaper area, (c) moves involving a larger mortgage than necessary to buy the new house, and (d) borrowing on the security of a house without purchase, sale or move. The first category, "last-time sales", includes sales by beneficiaries or executors, sales by (or on behalf of) home owners who have gone into residential care or moved in with relatives, sales by emigrants, sales by divorced couples who do not buy other houses or flats, or sales by couples where both previously owned homes. The key characteristic of this form of equity withdrawal is that it takes place through the sale of houses, so the amount is likely to vary strongly with the volume of sales and purchases. The second category, "trading down" by moving owner occupiers, involves buying a house for less than the sum received for the house being sold. Holmans observes that mortgage finance is not necessary for such moves. Sometimes, the mortgage will already be paid off, or the proceeds of the move enable the outstanding mortgage to be paid off. He notes that trading down moves can be combined with raising new funds by taking out a mortgage on the house being bought.

The third category, which we can term "over-mortgaging", involves taking out a larger mortgage than would be required if all the proceeds from the sale of the previous home were used to help finance the purchase. The defining characteristic of this type of equity extraction is that it can only take place in the course of a move of house in which the previous residence is sold and a fresh one is bought; and it requires mortgage finance. The fourth category, "re-mortgaging" or borrowing on the security of a house without a purchase, sale or move taking place, can involve a further advance on an existing house, replacing existing mortgages by new and larger ones, or raising a fresh loan alongside an existing one or on an unmortgaged house. The crucial distinction, according to Holmans, is that whereas the first three categories depend on the purchase and sale of dwellings and are therefore related to house prices and the total volume of transactions, the final category is independent of the volume of sales. There are no readily available figures on this category and Holmans estimates it as a residual, with all the consequent problems of measurement error.

The four categories of housing equity withdrawal (HEW) identified by Holmans comprise total gross housing equity withdrawal. Net housing equity withdrawal comprises gross HEW net of injections of equity which either add to the value of the owner occupied housing stock without any corresponding increase in debt secured on it or reduce the indebtedness secured on the stock. The former include expenditure on improvement of owner occupied dwellings paid for by owners themselves, rather than by borrowing; first-time buyers' deposits; and first-time purchases without a loan.

Holmans's findings show first that net housing equity withdrawal has tended to be cyclical, with peaks at the peak of each housing market boom. Net HEW (which was negative in 1970) rose to £163 million in current prices in 1971 and £608 million in 1972 before falling back to £174 million in 1973 and –£37 million in 1974. It then rose steadily from 1975 to 1977, reaching a peak of £1,070 million in 1978, and fell back in 1979 and 1980. It then began to grow rapidly, particularly in 1982 when it rose to £4,039 million, rising to a peak of £6.17 billion in 1988, falling back to £11 billion in 1989 and 1990. The pattern has been one of cyclical fluctuations on a rising long-term trend (Table 6.6).

Table 6.6 Net housing equity withdrawal derived from net lending for house purchase and net capital expenditure on owner-occupied dwellings (£m)

	Capital expenditure				Net lending	Net equity withdrawal
	New dwellings	Improvement	Net purchases	Total		
1970	931	197	89	1,217	1,190	−27
1971	1,134	308	160	1,602	1,765	163
1972	1,429	332	321	2,082	2,690	608
1973	1,789	508	259	2,556	2,730	174
1974	1,575	438	146	2,159	2,122	−37
1975	1,893	730	195	2,818	3,361	543
1976	2,074	825	185	3,084	3,758	674
1977	2,161	862	267	3,290	4,140	850
1978	2,685	1,002	466	4,153	5,223	1,070
1979	3,105	1,680	612	5,397	6,142	745
1980	3,233	2,316	1,020	6,569	6,976	407
1981	3,302	2,489	1,606	7,397	8,983	1,586
1982	4,302	2,528	2,822	9,652	13,691	4,039
1983	5,347	2,265	2,643	10,255	14,466	4,211
1984	6,042	2,841	2,388	11,271	17,151	5,880
1985	6,519	3,934	2,379	12,832	18,965	6,133
1986	8,224	4,539	2,716	15,479	26,845	11,366
1987	9,942	5,212	2,885	18,039	28,802	10,763
1988	13,371	6,127	3,677	23,175	39,343	16,168
1989	11,940	6,682	3,424	22,046	33,195	11,149
1990	10,727	6,172	3,977	20,876	32,600	11,724

Source: Holmans (1991)

Equity withdrawal really began to expand in 1982 when the banks entered the mortgage market in a big way and lender competition replaced mortgage rationing. In current prices, net equity withdrawal had its biggest proportionate rise between 1980 and 1981, when it rose from £407 million to £1,586 million (+290%) and between 1981 and 1982, when it rose to £4,039 million (+155%). Although the absolute increase was much greater in later years, the proportionate rise was much smaller. There were, however, big rises between 1985 and 1986 (+85%) and between 1987 and 1988 (+50%) when the 1980s boom was at its height.

The cyclical boom in HEW and the massive expansion in the early 1980s is even more marked when net HEW is measured in 1985 prices (Figure 6.2). These figures show increases in the early 1970s from −£245 million in 1970 to £547 million in 1971 and £2,296 million in 1973, falling to £426 million in 1974 and −£228 million in 1975. In the early 1980s the growth is even more remarkable, rising from £90 million in 1980 to £1,460 million in 1981 and £4,064 million in 1982. Once again, the all-time peak was in 1988, with £12,277

139

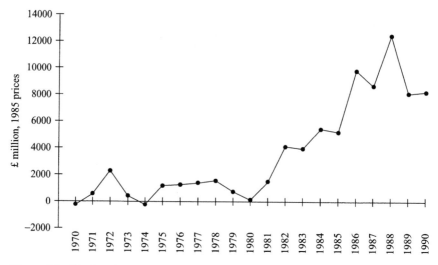

Figure 6.2 Net housing equity withdrawal, 1978–90
Source: Holmans (1991)

million. These figures very strongly suggest that the entry of the clearing banks into the mortgage market in the early 1980s was accompanied by a rapid surge in equity extraction as mortgages became easily available.

The third point to note is that equity extraction has not completely dried up since the housing market slump. It has merely fallen back in real terms to the level prevailing in 1986–87 before the peak of the boom. Holmans attributes this partly to the fact that the scale of housing equity extraction by last-time sellers and owners trading down is partly related to the growth of home owner-ship and higher house prices. Thus, although the volume of sales fell by 35% from its peak of 1.99 million a year in 1988 to 1.28 million in 1990, there was still a substantial level of sales and equity extraction. People still die and leave their houses, slump or no slump. It should be noted that the value of housing in estates passing at death in 1992–93 was £8 billion and, when allowance is made for the estates passing between spouses, the total value was still £5.6 billion.

The fourth and final point to note is that there has been a very marked shift in the internal composition of equity extraction during the twenty-year period 1970–1990. The proportion of gross equity accounted for by elderly last-time sellers fell from 52% in 1970 to 42% in 1979, to 30% in 1985, and an all-time low of 21% in 1990. This may be an anomaly, but even excluding 1990, the proportion fell to under 30% in the late 1980s. The proportion of gross HEW accounted for by other last-time sellers has been more variable, but it shows a secular decline from an average of about 18% between 1970 and 1981 to between 10% and 15%, with a downward trend to 10%, in the 1980s. The importance of these forms of HEW has fallen from about 67% in 1970–73 to 40% in the late 1980s and 30% in 1990. The importance of trading down has also fallen consistently,

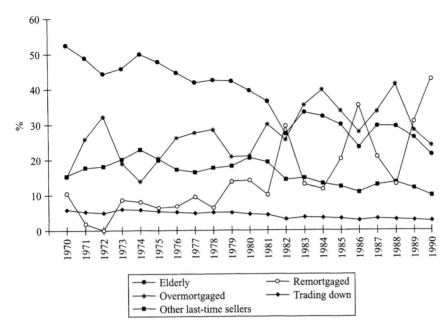

Figure 6.3 Gross equity extraction by type
Source: Holmans (1991)

but from only 6% to about 2.5%. This confirms the opening up of other possible forms of equity extraction since the early 1970s (Figure 6.3 and Table 6.7).

As the three "traditional" forms of HEW have declined in importance, so overmortgaging and remortgaging have grown in importance. The importance of overmortgaging is highly cyclical, with a peak in each of the three housing booms. It rose from 15% in 1970 to 26% in 1971 and 32% in 1972 before falling once again to 19% in 1973 and 13% in 1974. It then rose again to a peak of 28% in 1978, fell back to 20% in 1979 and 1980, before rising steadily through the 1980s to a peak of 41% in 1988. This form of HEW thus appears to be very strongly correlated with housing booms. There is a growth in turnover and mortgages may be easier to obtain. The remortgage and further advance category is a residual one, as noted earlier. It is thus much more prone to measurement errors than the other categories. Nonetheless, while it was stable at around 7% from 1973 to 1978 it experienced a dramatic, though very variable, increase during the 1980s, rising to a remarkable 30% in 1989 and 42% in 1990. On closer inspection however, it appears that this form of HEW may be counter cyclical. It was 10% in 1970 but it fell to zero in 1972 at the peak of the boom. It then rose to 14% in both 1979 and 1980 when the late 1970s boom had already peaked. It rose to 35% in 1986, but fell to 12% at the peak of the boom in 1988 before rising again as the market slid into recession.

Table 6.7   Summary of gross housing equity withdrawal (£m)

	Elderly	Other last-time sales	No mortgage trading down	"Over-mortgaging"	Re-mortgages further advances	Total
1970	482	143	53	140	96	914
1971	651	238	70	346	22	1,327
1972	869	357	97	638	-4	1,957
1973	850	375	111	358	161	1,855
1974	865	400	97	235	139	1,736
1975	1,333	572	146	556	178	2,785
1976	1,437	560	161	844	215	3,217
1977	1,565	622	178	1,040	355	3,760
1978	1,800	758	208	1,220	266	4,252
1979	2,094	903	240	1,040	686	4,963
1980	2,248	1,175	259	1,191	805	5,678
1981	2,678	1,421	303	2,198	729	7,329
1982	2,962	1,563	340	2,793	3,212	10,870
1983	3,818	1,689	401	4,034	1,455	11,397
1984	4,449	1,825	466	5,426	1,587	13,753
1985	4,743	1,945	513	5,356	3,212	15,769
1986	5,334	2,419	607	6,325	8,012	22,697
1987	6,774	2,926	731	7,694	4,752	22,877
1988	8,832	4,070	886	12,326	3,852	29,966
1989	6,617	2,957	689	7,089	7,661	25,013
1990	5,879	2,676	700	6,614	11,770	27,639

Source: Holmans (1991)

While it is dangerous to read too much into these figures because of their residual character, it is suggested that this form of HEW may increase in importance during the market downturns when sales fall and it is difficult to extract equity through moving house. The sharp increase in the late 1980s may reflect owners trying to refinance high interest debt on credit cards and overdrafts with lower rate long-term mortgage debt. If this is correct, it may also help explain why equity extraction has not fallen as much as might be expected with the housing market slump.

Although the proportion of gross HEW accounted for by last-time sellers has steadily declined over the 1970s and 1980s, its absolute value has risen steadily: from £482 million in 1970 to £8,832 million in 1988 (at current prices), falling back slightly to £5,879 million in 1990. The point, of course, is that the absolute size of the overmortgaging and remortgaging categories has risen much faster. Overmortgaging rose from just £140 million in 1970 to around £1 billion in 1977–79. It doubled between 1980 and 1981 and then rose sharply in 1988 to a peak of £12,326 billion. But the proportion of equity released by last-time sellers fell from an average of 70% in 1970–80 to 51% in 1981–84 and 42% in 1985–90. The balance appears to have been taken up by an upsurge of "overmortgaging" and remortgages. This has been particularly marked since 1989. The volume of equity release from last-time sellers fell from £11.2 billion in 1988 to £6.7 billion in 1990, and "overmortgaging" by moving owners was halved from £10.7 billion to £5.2 billion, while remortgages and further advances are estimated to have risen from £3.4 billion to £9.2 billion over the same period. Measured in 1985 prices gross equity withdrawal rose from a stable £7–8 billion a year from 1972 to 1980 to £12.6 billion in 1982 and 1983. The level moved up a gear again in 1986 when it reached £21.7 billion and rose to a peak of £26 billion, before falling back to about £21 billion in 1989 and 1990.

Overall, Holmans's data confirm the dramatic transformation which has taken place in equity extraction since the early 1970s and particularly during the 1980s. Whereas most home owners were traditionally limited by the availability of mortgage finance to some equity extraction on moving, trading down prior to death and leaving their property to beneficiaries, it is clear that this twenty-year period has seen the rapid expansion of other forms of equity extraction. Work by Westaway (1994) suggested that the level of housing equity withdrawal would recover sharply in the mid 1990s to reach a similar level in 1995 to that of the late 1980s; that it would continue to grow into the early years of the next century, accounting for between 3% and 4% of personal income; and that the "quasi consumer credit" element of equity withdrawal will account for around 4% of consumer income in the early years of the next century as inheritance increases.

What are the macro-economic implications of Holmans's analysis? Although the level of equity extraction should pick up when the housing market revives, there may not be the sudden upsurge some expect. First, potential equity extractors may simply switch back from further advances and additional mortgages to overmortgaging through sales. Second, lenders are likely to be less generous in their lending. They have burnt their fingers badly once and are unlikely to want to do so again. The surge in equity extraction in the early 1980s may have been a one-off adaptation to the greater availability of mortgages. Home owners may also be more wary in the 1990s of swapping equity for debt. Overall, it may be a mistake to look to the housing market as a way of kickstarting the economy. These questions are discussed in more detail in Chapter 8, which focuses on the housing market and the wider economy.

CHAPTER 7

Housing Careers and Housing Investment Strategies

Previous chapters outlined the widely held views that housing is, or was, a good investment, and summarised the evidence on the rates of return derived from housing. Chapters 4 and 5 analysed the distribution of gains and losses and housing wealth. But the change from long-term house price inflation to a more uncertain regime raises questions regarding the attitudes and behaviour of home owners towards ownership as an investment, which are the subject of this chapter.

Housing is an unusual commodity in that it possesses both a use value and an exchange value. It functions both as a home and a roof over people's heads and as a financial asset: a hedge against inflation and a store of wealth, and a source of both capital gains and losses. This dual role of housing, which is shared by certain other commodities such as antiques, fine art, classic cars and the like, poses a set of interesting and important questions regarding home owners' attitudes, motives and behaviour. Put simply, the key issue is the extent to which the behaviour of some or all home owners and potential owners is primarily motivated by use value considerations (house as home), or exchange value/ investment considerations (houses as financial assets) or by some mixture of the two, and the extent to which the mix of motives may have changed over time. Do owners buy houses primarily as places in which to live, moving house in response to their changing employment, household and lifestyle requirements, or do they buy houses primarily to maximise their potential capital gains (or to minimise the financial risks)? In many respects, buying a house entails major decisions about capital investment even though households may not see it as such. By taking out a mortgage, households are committing themselves to regular repayments over a long period of time and, when they put down a deposit, it represents investment or consumer spending foregone, irrespective of whether buyers are aware of this or not.

Housing: Investment or Consumption Good?

There has been considerable debate in the housing literature about the extent to which home ownership functions as an investment or as a consumption good or both. Traditionally, home ownership tended to be treated solely in terms of its use value by sociologists and social policy specialists who were primarily concerned with questions of living and space standards, access to housing and social differentiation of the housing market. Economists, on the other hand, have traditionally tended to analyse home ownership in financial terms, even though they recognised it had both a use and an exchange value. Analyses focused on the user costs of housing and the stream of benefits to its occupants, as well as the nature of housing as an asset.

Since the 1970s, however, a variety of authors have recognised that long-term house price inflation and the growing significance of capital gains and housing wealth may have changed the way in which housing is viewed. Sternlieb and Hughes (1972) stated that America had become a "post-shelter" society for many households, asserting that "For all but the most affluent in our society, a house is not merely a home, it is typically a major repository of capital investment and stored equity . . . Increasingly . . . houses are purchased to be sold not to be lived in." Similarly, Anthony Downs (1980: 2) suggested in a perceptive early article "Too much capital for housing?" that the attitude of American consumers towards housing has fundamentally changed: "Housing is no longer considered merely shelter. Many buyers now view it primarily as an investment that allows them to accumulate capital and to hedge against inflation." He argued that in addition to the tax advantages offered by home ownership, such as the exemption from capital gains and the ability to offset mortgage interest payments against taxable income, inflation magnifies the other benefits of investment in housing through the leverage that is possible from a small deposit. Downs went on to argue that:

> Consumers are alert to this opportunity. Millions of households have rushed to buy homes, thereby stimulating house price increases at a rate greater than the overall rate of inflation. . . . Ironically, rising house prices have not curtailed the number of units demanded, as economic theory would predict. Rather they have increased the number by stimulating greater total demand by people who expect additional price increases.

David Thorns (1981a: 707) also suggested from an Australasian perspective that:

> the increasing value of housing since the 1960s has had an effect in developing people's awareness of the exchange value of property as well as its use value. Thus people began to consciously develop their housing as a form of capital gain which they could reap through a house sale and either develop to further housing ventures or diversify into other investment areas.

And in a British context, Murie and Forrest (1980: 3) stated that:

> the role of housing as a source and store of wealth . . . influences individual household behaviour. People buy houses not only for use, but also as an investment to increase their personal assets and acquire wealth. Housing choice and decisions to change dwellings are not just decisions about consumption and use . . . They involve decisions about investment,

about using income available once essential needs have been met, about maximising net income and about returns on savings . . . This element in housing decisions is not always separable from other elements.

The extent to which people behave as Sternlieb and Hughes, Downs, Thorns, and Forrest and Murie suggest is a problematic question which has been the subject of a great deal of heated academic debate. This chapter looks at the extent to which home owners view house purchase primarily as financial investment and engage in a variety of strategies to maximise capital gains, or whether homes are bought mainly as somewhere to live or as some mix of the two motives. There are big differences between the claims of Sternlieb and Downs that housing is viewed primarily as an investment and the suggestion of both Thorns and Forrest and Murie that housing now functions in a complex and variable mixture both for use and investment.

There was a strongly held belief in the press in the mid to late 1980s that many households, particularly in the South East, were behaving speculatively and attempting to maximise capital gains from the housing market by policies of trading up, buying expensive properties, properties in up and coming areas or similar strategies. There a good deal of anecdotal evidence of this. In their book *Voices from the Middle Classes* Deverson and Lindsay (1975) quote two home owners as saying: "I'm not the least bit interested in houses as such and I'd be quite happy living in a tent, but I'm excited by the financial possibilities of houses. I regard it as a form of earning because I don't work", and "We're having the kitchen and breakfast room bashed down to make a super, incredibly flash kitchen, and we're doing the same with the bathroom and loos. It will cost £3,000 and as soon as it's done it will go on the market."

Almost every home owner knows people who have acted in this way, buying houses primarily for renovation and resale: some even make a living from it – they have become part-time property developers who live in the house while the work is being done. Farmer and Barrell (1981) suggested that, given the large capital gains from home ownership in Britain in the 1970s, some home owners might be pursuing an entrepreneurial strategy, consciously maximising their housing investment and trading up to maximise potential capital gains. They did not test this thesis in any systematic empirical way, however, and for the most part the debate has proceeded solely on the basis of assertion, surmise and theoretical speculation.

Some comparative research on the motives and attitudes of home owners was done by John Agnew (1981) in Leicester and Dayton, Ohio in the mid 1970s. He found that 85% of his sample of owners in Leicester said making a profit from their house was not an important consideration compared to just 27% in the USA. None of his Leicester sample viewed profit as a very important consideration, and 43% saw it as totally unimportant compared to just 3% in Dayton. Couper and Brindley (1975) also found that only 25% of home owners in Bath viewed their home primarily as a financial asset and an investment. These findings

Table 7.1 The importance of profit to owners in Leicester and Dayton, 1975 (%)

	Very important	Important	Unimportant	Very unimportant
Leicester	0	14	42	43
Dayton	13	60	24	3

Source: Agnew (1981)

may reflect a pre-inflationary attitude to home ownership in Britain, but they are nonetheless important (Table 7.1).

Peter Saunders (1990), one of the leading proponents of the investment view of housing, undertook a survey of just over 500 home owners in the three towns of Slough, Derby and Burnley in 1986. In response to questions about why they had decided to buy in the first place 29% of owners said that by buying they could get something back for the money they would pay out in rent and 20% of owners expressed an investment view of housing. In response to a question concerning the advantages of owning versus renting, 38% said that ownership represents a stake in an appreciating asset. When asked if they had made money out of housing, 34% said yes, 11% said no, 27% said that they had made a paper gain which could not be realised, and 19% said that they would have to move house to realise any gains. Saunders noted that 13% of his respondents not only said that they had made money, but outlined how they planned to cash in on their gains. He argued that:

> Many home owners bought housing because it seemed a good investment, see asset appreciation as a major advantage of owner occupation, and believe that they have accumulated real gains as a result of buying a house. It would be strange if it were not so, for if people have been accumulating capital at the rate that we have demonstrated ... they would need to be extremely unobservant not to have tumbled to the fact. (1990: 32)

There can, I think, be no doubt about the validity of these findings, not least because Saunders undertook his research in three towns with large working-class populations, not in prosperous middle-class areas of the South East where an investment view of housing might be expected to be much greater. But Saunders also suggested many households are explicitly pursuing entrepreneurial housing careers and he argued that:

> The fact that so many owner occupiers now move so frequently, and yet move such small distances and without any employment related motiva-tion, can *only* be explained by the fact that many of them are following a deliberate and coherent investment strategy through the housing market ... Not only, therefore, do home owners make real gains out of their

housing, but they are aware of this and many develop strategies designed to maximise it. (1990: 200, emphases added)

This is a quite fallacious argument. Although many home owners do move frequently, and for non-employment-related reasons, it does not follow that the explanation is that they are pursuing an investment strategy, coherent or otherwise. People move house for many reasons, including changes in household circumstances such as marriage, divorce or children, and many move simply to improve their housing circumstances (for a bigger or better house, or to a nicer area). Indeed, as Saunders himself points out: "In our survey, 38% of all households had definite plans to move house in the future." Of them, 18% intended to move to a better house, 18% planned to move to a better neighbourhood, and 12% simply intended to move "up-market" (1990: 200).

These are not investment motives but use-related ones. The desire to maximise housing consumption is probably a dominant factor. As Savage et al. (1992: 92) note, "an investment view is not always commensurate with investment behaviour . . . entrepreneurial housing strategies might be as much use value directed as investment directed". They concluded that "there is little evidence that people pursue housing careers in the sense of planned, long term trajectories related to maximising financial gains" (1992: 157). In their study of housing careers in two different areas of Bristol, an affluent middle-class area and a working-class area, Forrest and Murie (1987b: 336) asked whether certain households "actively manipulate the housing market to climb the housing ladder . . . making decisions about when to move and what to buy related to obtaining the best return through housing transactions". They found that in the working-class area "financial strategies tend to revolve around minimizing indebtedness and escaping from private landlordism" and that "even among those in the affluent area who had made a rapid ascent of the housing market there was little evidence to indicate a determined and conscious pursuit of investment and speculation in housing" (1989: 84). They accept that the investment potential of housing is important, but they point out that it is necessary:

to distinguish the factors which influence specific housing decisions from the primary determinants of residential movement or nonmovement. Decisions to move (or not to move) may not be investment driven, but financial/ investment factors coalesce with other considerations such as house size and location, education and social and physical environment to shape housing histories. . . . The view that housing investment is a conscious element in most movement decisions by home owners is not in dispute. What is at issue is how important an aspect it is. (1989: 84)

Similarly, Forrest et al. (1990: 155) argue that Farmer and Barrell's picture of housing entrepreneurs "suggests that it is an exotic group with secure jobs rather than a large sector of home owners. Attempts to identify the group in practice suggest they are a minority protected species."

Motives for House Purchase in the South East

To test Saunders's thesis MORI was commissioned to ask 1,000 home owners in five locations in the South East a number of different questions concerning attitudes to home ownership and why they had bought their current home. The responses clearly refute Saunders's claims. The most important reasons given were "bigger/better home" (29%), and other housing reasons (29%), followed by better neighbourhood (18%), change of job/to be nearer job (17%), and other non-housing reasons (13%), wanted garden (9%), better value for the same cost (8%), change in family circumstances (8%), to be nearer family/friends (7%), different kind of environment (7%) and new relationship/marriage (5%). "To increase investment in housing" scored only 5% and "potential for capital gains" just 3%: the same as "to reduce housing costs" (5%). When asked "what was *the most important single reason* for moving", 17% of respondents noted "bigger/ better home", 20% "other housing reasons", and 13% "change of job" against just 2% who identified an "increase in housing investment" and 1% who identified "potential for capital gains". Maximising financial gains are clearly not a major reason in deciding to move house.

Secondly, we asked home owners which of a series of statements *best described* their attitude to financial factors when they bought their current home. The statements, and proportions of respondents (in rank order), were:

(1) Dwelling/area important, financial factors not major – 33%
(2) Wanted somewhere that would hold/protect investment – 20%
(3) Dwelling/area was important, not financial factors at all – 16%
(4) Wanted the cheapest property we could afford – 12%
(5) Possibility of financial gains a major factor – 4%
(6) Wanted to limit costs/payments on the property – 4%
(7) Wanted to maximise capital accumulation – 2%
(8) None of these – 7%

Again, it is clear from the responses that, although financial factors are important, gains are not the most important factor. The largest response – 33% – was to statement number 1, that it was the dwelling and area that were of primary importance and financial factors were not major, followed by what can be described as a "defensive" financial approach of buying a home that would hold its value and protect investment – 20%, followed by dwelling/area important, financial factors not at all – 16%. Only 4% gave financial gains as a major factor and just 2% wanted to maximise capital accumulation compared to 12% who stressed affordability: the cheapest property they could afford.

We also asked home owners to agree or disagree with several statements. The first was: "It's important to buy the most expensive house possible so as to maximise our long term capital gains." Only 3% strongly agreed with the statement, 14% tended to agree and 12% neither agreed nor disagreed. By contrast, 40% tended to disagree and 27% strongly disagreed. In other words, two-thirds disagreed to some extent and just 17% agreed. When we cross-tabulated the data with date of purchase we found agreement decreased with recency of purchase and there was very strong disagreement (over 30%) from those who bought in the 1980s, particularly the late 1980s. The proportion of the sample agreeing with the statement increased with social class and with the number of houses owned: of those who had owned one or two houses 14% agreed compared to 26% who had owned three or more houses. Significantly, the highest proportion agreeing with the statement (31%), was found in the expensive Chiltern area, which has the highest proportion of multiple home owners and high social class, and lowest in Haringey and Milton Keynes (11%): both areas characterised by high proportions of first-time buyers, cheaper properties and recent buyers. Similarly, the proportion who agreed with the statement rose with income: from 9% of those earning under £100 a week to 30% of those earning over £625 a week. Overall, higher-class, high-income people were most likely to agree that buying the most expensive home to maximise capital gains is important, but they are still in a minority. Interestingly, when we asked owners whether they agreed that the long-term trend for prices is always up and houses are a good long-term investment, we found that 15% strongly agreed, 46% tended to agree, 18% disagreed and only 5% stongly disagreed. Notwithstanding the slump, the majority of owners were positive about long-term house price inflation.

Significantly, the proportion agreeing with the statement was higher in both Chiltern (71%) and Oxford (70%) than in Haringey (44%) or Milton Keynes (58%), which had a higher proportion of more recent first-time buyers. Once again, the proportion agreeing with the statement was much higher for those who bought many years ago (1959–65) falling to 54% of those who bought in 1980–84 and 49% of those who bought in 1985–88. There was no pattern of agreement by class, but agreement was strongly related to current house price: 56% of those with houses worth less than £60,000 agreed compared to 67% of those with houses worth over £160,000. The results indicate that the longer one has owned, and the more expensive the house, the greater the belief that houses are a good investment: a reasonable view all considered.

Like Saunders, we also asked owners what motivated them to buy initially, giving them a wide range of options. By far the most important reason, given by 31% of the respondents, was "Long term, owning is better value than renting", followed by "Greater security and control over housing" (17%) and "Home ownership is a good way to accumulate capital" (12%). "Got married" (9%) was the next most common reason followed by a variety of reasons (wanted to rent, but no suitable accommodation, 7%; would never be able to own if didn't buy, 7%; wanted to be financially independent after retirement, 7%; wanted to start a family/had young children, 6%; thought house prices cheap at the time, 6%; greater choice/better value, 6%; wanted to live in area of my choice, 5%). This suggests that family- and choice-related reasons form an important secondary motivation for buying. Interestingly, the proportion who listed "long term, owning is better value than renting" as a reason rose with recency of purchase, from

24% of those who had bought prior to 1969 to 42% of those who had bought after 1989. This suggests that, slump notwithstanding, younger, more recent buyers are still committed to the view that home ownership is better value than renting. In general it seems that the great majority of home buyers in the South East buy and move house primarily for housing, area and job related reasons, not for maximizing capital gain.

Housing Careers and Strategies

The term "housing career" is used to describe the sequence of moves made by households from one type of property to another and one area to another over the course of their lives. It should not be taken to imply an upward progression but rather a series of moves which may be up, down or horizontal; Jones (1978) found that moves downmarket to cheaper or smaller properties are quite common. In some cases housing careers will involve long-term immobility. It is, however, possible to identify a variety of different types of characteristic careers (Reidy, 1995) which all usually start in the parental home and involve various sequences of moves, perhaps into private renting and hence into owner occupation (Forrest and Kemeny, 1982) or council renting (Ineichen, 1981; Payne and Payne, 1977; Hamnett, 1984), or in some cases from the parental home into council renting and thence into owner occupation. Some people, as a result of marital break-up, repossession or unemployment, move back from owner occupation into council or private renting, but these moves are less common (Reidy, 1995). Forrest and Murie (1987b, 1989b) have also examined the sequences of moves within ownership, which in the case of highly mobile professional and managerial owners often involve moves upmarket from one area to another as people change jobs over the course of their labour market career. They found that working-class home owners tended to be less mobile and were more likely to move locally if at all.

In the sense of a series of related events the term "career" is a useful label, but to what extent do individuals have long-term expectations or aspirations for their housing careers and consciously attempt to achieve their aspirations by way of particular housing market or employment "strategies"? The concept of strategy has become increasingly popular in the 1980s and 1990s among sociologists, social historians, anthropologists and others concerned to understand social behaviour (Anderson et al., 1990; Crow, 1989). But there has been little interest until recently in the idea of housing strategy (Pickvance and Pickvance, 1994; Forrest and Kennett, 1996; Saunders, 1990). It would be surprising, however, if households simply made housing decisions on a completely *ad hoc* or even random basis. On the contrary, there is a strong likelihood that for some owners housing decisions form part of a conscious long-term strategy to achieve a given, or changing, set of housing goals. Such goals may be to do with location, house type, size, garden size or other consumption factors, or they may be more or less explicitly financial: such as buying a house that will continue to appreciate in value and will, when the mortgage is paid off, yield sufficient equity to provide care in old age or to supplement a pension. Anecdotal evidence suggests this type of motive is quite common amongst self-employed home owners, who may have only limited pension provision. Other home owners may simply want the security of a substantial asset with no specific goal in mind, and yet others may approach home ownership in an explicitly entrepreneurial way, buying and renovating houses to resell at a profit. Many first-time buyers, on the other hand, will be primarily concerned with trying to buy the best possible home at a minimum or affordable price. What is certain is that, insofar as people have explicit or implicit strategies, these are likely to involve a projected series of moves to attain the desired objectives. One way of looking at these sequences is in terms of moves up and down "housing ladders", and this concept is discussed below.

Snakes and Ladders in the British Home Ownership Market

In western capitalist countries the home ownership market is differentiated by the type, size, location and price of properties. The bottom end of the home ownership market in Britain consists of small flats or terraced houses in less attractive areas, whereas the top end of the market consists of larger, more expensive, better quality flats and houses in more attractive locations. Because of the competitive nature of the market, and the allocation by price and ability to pay, there is generally a strong positive relationship between type and price of property and class and income (Forrest et al., 1990). This relationship is not static however, and there is frequently a progression over the course of a household's "housing career" from smaller, less expensive properties, to larger and usually more expensive properties as income rises or as the household grows. In Britain, the most common metaphor for this process is that of climbing the housing ladder. Households are seen to "get a foot on the ladder" by buying a starter home, perhaps a small flat, an inner city terrace or a small modern "box" on a suburban estate and subsequently "trading up" as circumstances permit, perhaps making a series of moves into progressively more attractive and expensive property.

The phenomenon of "trading up" is not unique to Britain, but is particularly marked, not least because the majority of first-time buyers buy in their twenties, and can only afford a relatively inexpensive house. In other countries, such as Germany, where private renting is more easily available, ownership is very expensive, and mortgage interest tax relief is only given on the first house; new households tend to rent, only buying later in life and remaining in the same house for many years (Muellbauer, 1991). The process is not, of course, one-way, and some households also trade down, particularly in the latter stage of the household life cycle when children may have left home or when people retire and are looking for a smaller or less expensive home. There is also a degree of trading down associated with divorce and enforced sales (Kendig, 1984a; Symonds, 1989).

The notion of an owner occupied housing ladder is widely held in Britain. The term is loosely used but essentially the different types, sizes and quality of housing are seen as forming a ladder of desirability with the rungs of the ladder being differentiated by type, size, location and desirability, which are reflected in the price of property. The structure of the ladder is thus defined by the valuation individuals place upon different types of property and the degree of competition for them. Smaller, cheaper or less desirable properties comprise the bottom end of the ladder and highly desirable properties which attract high prices comprise the top rungs of the ladder. The structure of the ladder is not fixed. It will reflect the changing evaluation of and structure of demand for properties of different types and locations. Movement up and down the ladder takes place over the life of individual owners as they seek to match their aspirations with their circumstances and resources. As the Nationwide Building Society put it in 1976:

It is well known that people tend to move up the housing ladder. Most young people buy their first home in their 20's (about 70% of first time buyers are under 30) usually after a short period of living with in-laws or in rented accommodation. Considerably less than a quarter of first time buyers purchase new property and well over a quarter purchase property built before 1919. First time buyers tend to trade up to a better and usually larger house in their mid 30's, although moves of this kind are spread over the 25–44 age group. There is a clear preference for detached houses with gardens, which form about a third of their purchases.

The housing ladder should be seen probabilistically, in terms of general tendencies and movement patterns, rather than absolute or determinate regularities, and there is no assumption that all home owners will necessarily move up the ladder or that they will move up equally. But Forrest et al. (1990) are critical of what they see as the implicit assumptions underlying the metaphor of a home ownership ladder. They argue (1990: 28) that:

The notion of a starter home, for example, contains an implicit assumption of an inevitable movement up the rungs of the home ownership ladder. But this is by no means assured, nor can we assume that all home owners have the same set of aspirations. Everyday experience suggests that different groups are on very different ladders, or at least if there is one ladder it is almost infinitely extendable.

They also assert (1990: 101) that:

If the owner occupied market is represented as a ladder, it is assumed that the rungs are on the same ladder, with the promise, if not the possibility, of progression from the bottom to the top. There may be winners and losers, but this is presented as a matter of luck and housing market acumen, rather than as a product of the odds being systematically biased in favour of some at the expense of others.

And they state (1990: 125) that:

There is not one housing ladder with a competitive scramble fuelled by a common set of motivations and values. Rather there are different ladders, which rarely touch or overlap. Some housing experiences do not involve more than one rung and some involve leaping to a higher rung without using those below. It is also inaccurate to assume that people necessarily evaluate their housing histories in terms of a series of escalating exchange values.

These critical observations have considerable force. It is clearly erroneous to assume that everybody has the same set of values and aspirations. Although some people may aspire to a detached house in its own grounds, by no means everybody would. Nor would everyone necessarily aspire to a large house or flat. In addition, the resources necessary to move up the ladder are very unequally distributed. The structure of the ownership market is more akin to a pyramid than a ladder with rungs of equal width. There are many terraced and semi-detached houses, but few detached houses in their own grounds, and it is clearly impossible for everybody to own a detached house even if they wanted to. Nor will all home owners progress up the ladder at an equal rate or from a similar starting point. Those with high incomes or wealth are likely to move directly to houses several rungs up on the scale of price and desirability, while others are hard pushed to get a foot on the bottom rung of the ladder and many make only one or two moves during the course of their ownership career, perhaps making little or no significant housing gain. Some people, of course, never move at all. We should also add that some people go down the ladder, sometimes on retirement or when children leave home, and sometimes because of divorce, unemployment or some other crisis. The implicit notion that most people continue to move up the ladder until they drop off the end at death rather in the manner of a Hieronymus Bosch picture is clearly highly unrealistic. The ladder is not one-way and it is possible to go down as well as up: sometimes rapidly and unpredictably. Doling, Ford and Stafford (1991: 115) state that mortgage arrears and repossessions are particularly prevalent among young, first-time buyers who borrow to the limit to get a foot on the ladder, and they note that "getting a foot on the ladder is no guarantee of not slipping off again". Jones (1978) found downward moves were common in the North West in the mid 1970s, although this may have been specific to that area and period.

The criticisms of the notion of a single housing ladder, up which progress is smooth, unproblematic and inevitable, have considerable strength. But few people would subscribe to such a restricted and mechanical conception of the home ownership ladder, and it is important that we do not allow these valid criticisms of the housing ladder to obscure the fact that many home owners do in fact move to bigger, better and more expensive homes over the course of their home ownership careers. What is clear, however, is that the possibility of such moves is strongly related to class, income and age. Finally, it needs to be stressed that there is not a single national ladder, but a large number of locally specific ladders, each with very different characteristics and opportunity structure. The structure of the housing ladder, and the type and price of different rungs, differ very considerably within the South East, particularly between inner London, which is dominated by flats and terraced houses, and the rest of the region, where semi-detached houses are common. This makes interpretation of changes in property type tricky, as a terraced home in parts of inner London may be a £500,000 period house at the top of the housing market, whereas a terraced home elsewhere may be a 2–3 bedroom house costing £60,000. They have the same label in common, but in other ways they are worlds apart.

Housing Ladders in Five Areas of South Eastern England

A systematic study of home ownership careers in five locations in the South East of England carried out by Hamnett and Seavers (1994b) found evidence of the existence of a "housing ladder", in terms of patterns of moves between houses of different types, sizes, prices and other characteristics. The logical starting point for a consideration of home owners' careers is an examination of the distribution of the stock, followed by an analysis of housing mobility. Only then can we look at movement up and down the ladder.

Of the sample as a whole 28% lived in a detached house or bungalow, 25% in semi-detached houses or bungalows, 32% in terraced houses, 10% in a purpose-built flat and 4% in a converted flat. The distribution of property types was very different between areas, however. Table 7.2 shows that in Chiltern, an affluent largely rural area in the Outer South East, almost two-thirds, 64%, of households lived in detached houses and a further 18% lived in semi-detached houses. In Milton Keynes (a new city built since 1970) the figures were 44% and 37% respectively. By comparison, in the inner London boroughs of Hammersmith and Haringey, only 2% of households lived in detached houses, whereas 53% lived in terraced houses. At the other end of the spectrum, 34% of households in Haringey and 27% in Hammersmith lived in flats (compared to 8% in Oxford, 3% in Chiltern and 1% in Milton Keynes). The proportion of households living in terraced houses and flats ranged from 86% in Haringey and 80% in Hammersmith to 43% in Oxford and 16% in Chiltern and Milton Keynes. These differences reflect the age and building history of the different areas and suggest that the structure of the housing ladders will vary sharply from area to area.

The distribution of property values was relatively evenly spread across the range of values, with a slight negative skew. Of the 852 respondents who were able or willing to estimate the current market value of their property, 43% put it

Table 7.2 Type of property by area (%)

	Chiltern	Hammersmith	Haringey	Oxford	Milton Keynes	All
Detached	64	1	2	19	44	28
Semi-det	18	19	12	37	37	25
Terraced	14	53	52	35	18	32
P/B flat	3	19	23	7	1	10
Conv. flat	–	8	11	1	–	4
Total	100	100	100	100	100	100

n = 972
Source: South East Homeowner's Survey (1993)

Table 7.3 The distribution of current market value by area (%)

Market value £000	Chiltern	Hammersmith	Haringey	Oxford	Milton Keynes	All
<50	0.5	1.7	8.4	3.1	17.8	7.2
50–59	4.8	1.7	14.0	11.0	24.5	12.3
60–69	5.8	5.3	12.9	27.0	13.9	13.3
70–79	6.3	3.5	11.2	22.1	8.6	10.6
80–89	7.9	7.0	11.8	12.9	7.8	9.5
90–99	9.0	8.8	5.6	3.1	5.8	6.3
100–119	7.9	16.7	7.9	2.5	6.7	7.7
120–149	11.6	14.0	9.6	3.7	4.3	8.2
150–199	20.6	21.1	12.4	6.1	5.8	12.6
200–299	14.8	17.5	3.9	6.7	2.9	8.5
300 +	10.0	2.6	2.2	1.2	1.9	3.6
Total	100	100	100	100	100	100

n = 852
Source: South East Homeowner's Survey (1993)

in the lowest four categories and 53% were in the bottom five categories. There was, however, a second peak in the price distribution at £150,000–199,000 and a quarter of respondents estimated the current market value of their properties at over £150,000. The distribution of values by location varied very considerably, however, as Table 7.3 shows. Milton Keynes was the cheapest area: 18% of respondents estimated the current value of their homes at under £50,000, and another 25% at £50,000–59,000. Over 56% of homes in Milton Keynes were valued at under £70,000. This compares to 41% in Oxford, 35% in Haringey, and just 11% in Chiltern and 9% in Hammersmith. The median price in Milton Keynes was £65,000 compared to £70,000 in Oxford, £80,000 in Haringey, £126,000 in Hammersmith and Fulham and £127,000 in Chiltern. No fewer than 65% of properties in Chiltern and 72% in Hammersmith were worth £100,000 or more. The proportion was in Haringey 36% but only 22% in Milton Keynes and 20% in Oxford. Not surprisingly, type of property is clearly related to current

157

Table 7.4 Mean house price by type and area (£000)

	Detached	Semi-det	Terraced	P/B	Conv.	Mean
Chilterns	206	99	81	55	–	161
Hammersmith	190	234	138	95	83	142
Haringey	267	146	105	62	61	99
Oxford	166	76	73	70	61	90
Milton Keynes	112	61	47	–	–	82

Source: South East Homeowner's Survey (1993)

Table 7.5 Number of rooms by property type (%)

	1–2	3	4	5	6	7	8	9+	Total
Detached	–	3.3	8.9	16.7	22.7	19.3	13.0	16.0	100
Semi-det	–	6.2	26.3	41.6	16.0	5.8	2.9	1.2	100
Terraced	2.3	6.6	32.9	32.6	12.5	9.2	1.0	3.0	100
P/B flat	16.8	50.5	24.2	6.3	2.1	–	–	–	100
Conv. flat	26.5	41.2	26.5	2.9	2.9	–	–	–	100
Total	3.3	11.8	23.0	26.5	14.7	9.9	4.7	6.0	100

n = 956
Source: South East Homeowner's Survey (1993)

value. Of those living in properties estimated to be worth £160,000 or more 61% lived in detached houses (though 22% also lived in terraced houses, principally in the inner London boroughs). Conversely, only 7% of those in property worth under £60,000 lived in detached houses compared to 20% in flats and 37% in terraced houses. The differences in the price structure of housing markets is shown by the fact that only 10% of the detached houses in Chiltern were valued at under £100,000 compared to 25% in Oxford and no less than 54% in Milton Keynes. In the latter, 30% of detached houses were valued at under £79,000 and 10% at under £59,000. These detached houses are, for the most part, small detached three-bedroom modern houses, rather than detached houses in their own grounds. Conversely, 87% of all terraced houses in Milton Keynes were valued at under £59,000 compared to just 7% in Hammersmith and 15% in Haringey. In Hammersmith, 80% of terraced houses were valued at over £100,000 and 44% at over £150,000. A house type which may be the bottom of the housing market in one area can be near the top in another. The pattern of house prices by type and area is given in Table 7.4. The average price for a semi-detached house in Hammersmith was £234,000 against only £61,000 in Milton Keynes: a ratio of almost 4:1.

The marked variations in the type and price of property were also linked to variations in size (measured in terms of number of rooms excluding kitchen and bathroom). Table 7.5 shows that, while 71% of detached houses had six or more

Table 7.6 Number of homes owned by study area (%)

Homes owned	Chiltern	Hammersmith	Haringey	Oxford	Milton Keynes	All	n
1	20	64	66	55	35	47	453
2	38	19	24	19	27	26	251
3	19	12	6	16	18	15	141
4	13	3	3	4	9	7	66
5+	10	2	1	6	11	5	50
All	100	100	100	100	100	100	961

Source: South East Homeowner's Survey (1993)

rooms, this fell to just 26% of semi-detached and terraced houses and to 2% of purpose-built and converted flats. By contrast, 67% of flats had three rooms or less and 90% had four rooms or less. There is a clear hierarchy of size and property type. But again, there are very considerable variations between areas. While most terraced homes (and many detached and semi-detached houses as well) in Milton Keynes had four to five rooms, many of the terraced homes in Haringey and Hammersmith had seven rooms or more. There is a sharp divide between a cramped modern estate house and many of the spacious period terraces of inner London.

There is a clear differentiation of property type, price and number of rooms, and the structure of the home ownership ladder varies by area. To show that there is movement up the home ownership ladder, we also need to show that there is degree of residential mobility over time. The Building Societies Association (BSA) have conducted regular surveys of home owners in Britain, and their findings show that the average length of residence of home owners was just under seven years in 1983, 1986 and 1989. This is an average, however, and in each survey about 47% of owners had lived in their current home for over ten years. Not surprisingly, length of residence varies with age. Whereas 31% of those aged 20–24 had lived in their current home under a year, this fell to 18% of 25–34 year olds, 7% of those aged 35–54 and just 1% of those aged 65 plus. Conversely, 70% of those aged 55 and over had lived in their home for over ten years. Residential mobility is strongly age-related, and younger groups make most rapid progress up the ownership ladder (CML, 1989). Another measure of housing mobility is the number of homes owned. Our survey of owners in the South East shows that, whereas 47% of owners have owned only one home, 26% had owned two homes, 15% three homes, and 5% had owned five or more homes. Overall, 87% of owners had owned three homes or less. But the proportions owning different numbers of homes vary very considerably by area (Table 7.6).

The proportion owning one home is far higher in the inner London boroughs of Hammersmith (64%) and Haringey (66%) than in Milton Keynes (35%) or Chiltern (20%). Conversely, the proportion owning three or more homes is far

Table 7.7 Number of homes owned by age of principal earner (%)

Homes owned	20–29	30–39	40–49	50–59	60–64	65–74	75+
1	83	42	31	48	43	49	62
2	12	32	33	19	30	21	20
3	5	15	21	16	12	12	6
4	–	7	7	8	6	8	9
5	–	3	5	5	4	4	2
6+	–	1	3	4	5	6	1
Total	100	100	100	100	100	100	100

Source: South East Homeowner's Survey (1993)

higher in Chiltern (42%), Milton Keynes (38%) and Oxford (26%) than Hammersmith (17%) or Haringey (10%). These figures confirm the role of inner London as a first-time buyer location. Chiltern, Milton Keynes, and much of the Outer South East, are destinations for those who have already made several moves. In part, this reflects the availability of different types of property in different areas. Housing markets are shaped by the structure of property type and prices and by the characteristics of those who live there.

These figures do not undermine the notion of a housing ladder however, not least because they represent a static cross-section of owners, both recent and long established. When the figures are disaggregated by date of first purchase the proportion of owners owning only one home falls from 98% of those who bought in 1989 or after to 63% of those who bought in 1985–88 and 38% of those who bought before 1960. Whilst by no means everybody moves, the likelihood of having owned more than one home rises with the number of years owned. Although 30% of single-home owners bought before 1974, no less than 60% had bought since 1980. This suggests that, other things being equal, a large proportion of current home owners will in due course, own a sequence of several homes.

The relationship between number of homes owned and age is shown in Table 7.7, and reflects the combination of cohort and age effects. The proportion of owners who have owned only one home falls dramatically to a low of 31% for the 40–49 age group, but rises steadily to a high of 62% for the 75+ age group. Conversely, the proportion who have owned three or more homes rises to a peak in the 40–49 age group. This is because this group first entered the home ownership market in the early 1970s when ownership opportunities had increased. Older age groups had less opportunity to purchase their own home in the 1950s and 1960s when ownership rates were much lower.

The Number of Homes Owned by Occupational Class and Income

The number of homes owned, and any movement up the owner occupied housing ladder, is not simply a product of age or number of years owned. On the

Table 7.8   Number of homes owned by occupational class of principal earner (%)

Homes owned	Prof.	Man.	ONM	Skilled	Semi	Unsk.
1	27	33	54	50	76	84
2	29	34	21	28	17	5
3	25	17	12	11	4	10
4	10	9	8	2	2	0
5+	9	7	5	9	1	0
All	100	100	100	100	100	100

Source: South East Homeowner's Survey (1993)

Table 7.9   Number of homes owned by household income (£ per week)

Homes owned	<99	100–224	225–349	350–474	475–624	625–749	>750
1	72	60	56	39	30	27	18
2	20	18	25	30	36	34	28
3	7	9	12	18	18	18	24
4	0	6	2	6	10	11	21
5+	0	7	5	7	6	10	9
All	100	100	100	100	100	100	100

Source: South East Homeowner's Survey (1993)

contrary, the number of homes owned is strongly related to both occupational class and household income. Table 7.8 shows that the proportion of owners who have owned only one home rises from 27% of professionals to 84% of the unskilled. The converse is also true: 19% of professionals have owned four or more homes, compared to just 3% of the partly skilled and none of the unskilled.

The number of homes owned also varies closely with household income. Whereas 72% of households with weekly incomes of £99 or under had owned only one home, and 20% of them had only owned two, the proportion of households owing only one home falls steadily in line with income, and only 18% of households with weekly incomes of £750 or more had owned only one home. Conversely, some 30% of those with incomes of over £750 a week had owned four or more homes compared to 13% of those with weekly incomes of £100–224 and none of those with incomes under £100.

Direct Measure of Movement up the Housing Ladder

The tables discussed above are very revealing and show that property type, size and price are strongly related to number of homes owned and time since first

Table 7.10 Current property type by first property type (2 homes only) (%)

First home	Current home					Total
	Detached	Semi-det	Terrace	P/B flat	Conv.	
Detached	13.2	1.6	0.8	2.0	–	17.6
Semi-det	12.0	12.0	2.0	1.2	–	27.2
Terraced	4.4	10.8	12.4	1.6	0.4	29.6
P/B flat	4.0	5.6	6.4	–	0.8	16.8
Conv. flat	–	1.2	6.0	1.2	0.4	8.8
Total	33.6	31.2	27.6	6.0	1.6	100

$n = 251$
Source: South East Homeowner's Survey (1993)

purchase. They also show that property type, size and price are very strongly related to class and income. But these are indirect measures of progress. To try to show directly the existence of a home ownership ladder, it is necessary to show the transitions in terms of property types, rooms and price between first, previous and current home. This exercise is only possible for those who have owned two or more homes. It would be expected that if there is progress up a home ownership ladder more people will move from flats to terraced houses and to semi-detached and detached houses than the other way, at least up until onset of old age. The same should be true of prices and number of rooms. More people should buy more expensive houses than cheaper ones when they move and more should move to bigger houses with more rooms than move to smaller homes with fewer rooms.

Taking the 251 households who have owned only two homes, Table 7.10 shows a significant upward shift in property type between first and current home. Looking at the column and row totals, 25% of first homes were flats compared to only 7.6% of current homes. Conversely, 33.6% of current homes are detached compared to 18% of first homes and 65% of current homes are detached or semi-detached, compared to 45% of first homes. Looking within the matrix of moves, of the 33% of households currently living in detached homes, 13% were initially in detached homes, 12% were in semi-detached, 4% in terraced and 4% in purpose-built flats. The proportion (52%), of upmarket moves (above the diagonal) was five times the proportion (10%) of downmarket moves (below the diagonal).

Of the 31% in semi-detached houses, 12% had lived first in semi-detached homes, whereas 11% were in terraced houses, and 7% were in flats. Only 1.6% had moved "down" from a detached to a semi-detached house. A similar pattern is found for terraced houses. Whereas 12.4% had moved "up" into terraced houses from flats, only 2.8% had moved down. The general pattern of property type mobility is upwards.

Table 7.11   Current property type by previous property type (3 homes) (%)

Previous home	Current home					Total
	Detached	Semi-det	Terrace	P/B	Conv.	
Detached	24.5	4.3	2.9	–	0.7	32.4
Semi-det	18.7	9.3	2.9	2.2	–	33.1
Terrace	7.2	10.1	4.3	–	0.7	22.3
P/B	0.7	2.2	2.2	2.2	–	7.3
Conv.	0.7	0.7	2.2	1.4	–	5.0
Total	51.8	26.6	14.5	5.8	1.4	100

n = 139
Source: South East Homeowner's Survey (1993)

Table 7.12   Current property type by previous property type (3+ homes) (%)

Previous home	Current home					Total
	Detached	Semi-det	Terrace	P/B	Conv.	
Detached	31.8	5.4	3.0	0.4	0.4	41.0
Semi-det	15.7	8.1	4.2	1.5	–	29.5
Terrace	5.4	6.9	5.7	0.4	0.8	19.2
P/B	1.5	1.5	2.3	1.2	–	6.5
Conv.	0.4	1.1	1.5	0.8	–	3.8
Total	54.8	23.0	16.7	4.3	1.2	100

n = 261
Source: South East Homeowner's Survey (1993)

This pattern of upwards movement is repeated for those who have owned three homes. Table 7.11 shows the proportionate distribution of property types over previous and current homes. Looking first at the row and column totals, we see that the proportion in detached houses rises from a third (32%) to just over a half (52%), while the proportion in terraced homes falls from 22% to 14% and the proportion in flats falls from 12% to 7%. The proportion of upmarket moves (below the diagonal) is 46.1%, 3.3 times the proportion of downmarket moves (13.7%). The ratio is far lower than for those who have owned only two homes (5.1:1), but this is explicable given that this group have owned three homes and have therefore made some progress up the ladder on their first move. Similar trends can be discerned when we look at the pattern of changes in the number of rooms and house price between previous and current property (not shown). The majority of owners move up to larger and more expensive properties.

This pattern of upwards movement between property types is repeated for those who have owned three or more homes. Table 7.12 shows the proportionate distribution of property types over previous and current homes. Looking at the

163

Table 7.13 Current property type by first property type (3+ homes) (%)

Previous home	Current home					Total
	Detached	Semi-det	Terrace	P/B	Conv.	
Detached	13.6	2.3	1.9	0.8	–	18.6
Semi-det	20.2	7.8	4.3	0.4	–	32.7
Terrace	15.2	7.0	6.6	0.8	0.8	30.4
P/B	6.6	2.7	1.6	1.6	0.4	12.9
Conv.	0.4	1.9	2.3	0.8	–	5.4
Total	56.0	21.7	16.7	4.4	1.2	100

$n = 257$
Source: South East Homeowner's Survey (1993)

row and column totals, we see that the proportion in detached houses rises from 41% to 55%, while the proportion in all other property types fell. The proportion of upward moves (all cells below the diagonal line from the top left-hand cell to the bottom right-hand cell) was 37% while the proportion of downward moves was 16%: and the ratio of upward to downward moves was 2.3:1. This ratio is far lower than the ratio for those who have owned only two homes (5:1), which is explicable given that those who have owned three homes or more will already have made considerable progress up the ladder on their first or subsequent move.

The structure of moves between first and current home is shown in Table 7.13. This shows that the proportion living in detached houses rose from 18.7% in first home to 56% in current home. This is a much greater increase than that between previous and current home. Conversely, the proportion living in terraced homes and flats fell from 49% of first homes to 22% of current homes. It seems that a great deal of the progress up the home ownership ladder is made on the first move, and the comparison of flows between previous and current home for those owning three and three or more homes confirms this. The progress up the ladder of property types between the previous and current home is much less for those owning more than three homes than for those owning three homes only. The proportion owning detached houses increases from 32% to 52% for those with three homes, whereas it rises only from 51% to 58% with those owning more than three homes. The more homes people have owned the further, in general, they have moved up the housing ladder at any given stage.

It is clear that there is a home ownership ladder in the South East or, to be more accurate, a set of different ladders in different local housing markets. The ladders are differentiated by type, price, size and desirability, and not surprisingly, position on the ladder and ability to move up are related to occupational status and income, and length of time in the ownership market. While some of the criticisms made of the indiscriminate use of the housing ladder metaphor and the assumption that all can climb up are valid, the existence of a ladder cannot be denied.

Conclusions

This chapter has looked at the claims and assertions made regarding the extent to which home owners in modern Britain are primarily guided in their home ownership behaviour by investment and capital gains considerations. It has been shown that, despite the plethora of competing assertions, there is little in the way of clear evidence. Although most home owners today are far more aware of the investment aspects of home ownership than they were twenty-five years ago, this does not necessarily imply that their behaviour is more determined by such considerations. Saunders's view that most moves are undertaken for reasons of investment and capital gains has been strongly rejected. Most moves are undertaken primarily for reasons related to housing and household, though this does not mean that owners are unaware of the financial implications of their housing decisions. It appears that the perceived long-term relative costs and benefits of home ownership versus renting, the perceived greater degree of security and control and perceived potential long-term capital appreciation are the most important reasons in deciding to buy a house initially, although relatively few home owners in the South East think that it is important to buy the most expensive house possible in order to maximise capital appreciation. Notwithstanding the slump, a majority of owners agreed that the long-term trend for prices is up and houses are a good long-term investment: a not unreasonable view.

The chapter has also discussed the ideas of housing careers, housing strategies and housing ladders. There is strong evidence among home owners in the South East that, notwithstanding criticisms of the concept, there are distinct housing ladders in the region, and that a significant proportion of owners do move upmarket to larger, more expensive, semi-detached and detached houses over the course of their housing careers, and that upmarket moves considerably exceed downmarket moves. Position in the housing market is strongly influenced by social class, income, number of homes owned, and the length of ownership, but mobility exists and housing ladders are not a myth.

CHAPTER 8

Home Ownership, Housing Wealth and the British Economy

The late 1980s and early 1990s have seen extensive debate in Britain on the role of house price inflation and housing wealth on consumer spending and the state of the economy as a whole (Boleat, 1994; Bowen, 1994; Clapham, 1996; Costello and Coles, 1991; Cutler, 1995; Ermisch, 1990; Meen, 1995; Miles, 1992a and b; Muellbauer, 1990c; Pannell, 1992). The debate revolves around the extent to which the housing market reinforced or generated fluctuations in the national economy during both booms and slumps. More specifically, it focuses on the relation between rising house prices and equity during the late 1980s, the increase in consumer spending between 1986 and 1988 and the associated fall in personal savings. The concern was reversed in the early 1990s, when debate centred on the links between falling house prices and sales, higher levels of mortgage debt, stagnant consumer spending, the rise in the saving ratio and the continuing recession. The debate is important, because it was commonly argued in the early 1990s that a recovery in the home ownership market was the key to wider economic recovery in Britain. As such, the home ownership market is now seen to play a central role in both economic growth and recession in Britain. This does not seem to be the case in other countries and it is important to examine the argument and its validity.

The idea that the housing market may play an important role in the wider economy is not new. House building accounts for a significant proportion of GDP and it has long functioned as a key indicator of economic conditions. When the house building industry booms, the economy also booms. But there is also a view that the construction industry functions as a Keynesian regulator, used by government for counter cyclical demand management; Harvey (1974) argued that housing policy is part of a wider government objective of orderly accumulation of capital, economic growth and social and political stability. It functions "to iron out cyclical swings in the economy at large by using the construction industry and the housing sector as a Keynesian regulator" (1974: 244). This argument was developed further by Checkoway (1980), Walker (1981) and Florida and Feldman (1988), regarding the key role of suburbanisation in American capitalist development. The regulation thesis advanced by Florida and Feldman argues that the suburban owner occupied single family house "opened up new markets for automobiles, home appliances and consumer products, as well as a wide range of public and private services" (1988: 197). As they put it: "US Fordism was inextricably tied to suburbanisation which enhanced consumer demand and set the preconditions for a temporary cycle of self-reinforcing growth" (1988: 188). They do not suggest however that the "suburban solution" (Walker, 1981) was the only possible solution. Rather:

> housing's crucial place in US Fordism was the product of unique historical conditions. While the productivity increases of Fordist production opened up a space for rising wages and mass consumption, the emergence of consumption patterns organised around suburbanisation was the result of a period of class formation, class conflict and attendant patterns of state intervention. (Florida and Feldman, 1988: 188)

Though the case for house construction as an important economic regulator is well established, this does not of course mean that it will necessarily be used as such. On the contrary, there is a strong case to be made that the Thatcher governments allowed new private house building to fall sharply in each of the last two recessions and that public house building was sharply cut back in 1979 and continued to fall during the early 1980s recession. Housing construction was not used as an economic regulator under Thatcherism. Instead, it was arguably sacrificed to the ideology of the market and if anything it functioned to reinforce booms and slumps in the wider economy.

Housing Wealth and Consumer Spending as a Cause of the 1980s Boom

The theoretical basis of the debate is simple. Basic neoclassical consumption theory postulates that as personal wealth rises so does consumer spending. But traditional models of consumer wealth and spending ignored the role of housing wealth (Davidson et al., 1978; Hendry, 1984) and failed to predict the big rise in consumer spending in Britain and the fall in the savings ratio in the late 1980s (Carruth and Henley, 1990). Lee and Robinson (1989a) state that: "It is no exaggeration to say that the consumption functions in the main macro-forecasting models have broken down". Consequently a number of economists in Britain explicitly addressed the role of housing in the economy (Ermisch, 1990; Muellbauer and Murphy, 1991), focusing particularly on the key role of housing equity and equity extraction in consumer spending.

As housing is the most important component of personal wealth in Britain it was noted that rising house prices had led to increases in personal wealth and that, as a result, many home owners felt able to spend more and reduce their savings, to extract equity from their homes, or to borrow against the increase in the value of assets. As consumer spending accounts for a large proportion of national expenditure, house price inflation was seen as a key determinant of demand and, hence, of the overall growth of the economy. The housing market is thus seen to play a leading role in economic growth. This view was clearly spelt out by numerous commentators at the time. Patrick Foley (1991b: 1) stated that:

The process by which the housing market influences the wider economy is straightforward. Housing accounts for the largest part of personal sector wealth. When house prices are rising fast . . . personal sector wealth is also rising fast, and this should encourage higher consumer spending, which is wealth dependent. The process of equity withdrawal, or at least that part of it, which is mortgage borrowing for purposes other than house purchase or improvement, is how this higher consumer spending is financed.

Bob Pannell (1992: 6–7) of the Council of Mortgage Lenders also argued that:

> with consumers accounting for two-thirds of all spending in the UK, the spending, saving and borrowing decisions of households have direct and pronounced effects on the well-being of the economy as a whole . . . The sharply rising value of dwellings owned by the personal sector was a major factor inflating the paper wealth of households, especially during the height of the property market in 1986–88 . . . The consequent increase in individuals' sense of well-being may have helped to increase the readiness of the personal sector to spend and to borrow, thereby reinforcing the whole process and stimulating overall economic activity.

Robin Leigh Pemberton (1986: 529), the then Governor of the Bank of England, argued that housing equity extraction may be linked to consumer spending: "The leaking of lending secured on a first mortgage, used for other purposes, may . . . play a significant role in fuelling the expansion of consumer spending, and the entry of the building societies into the consumer lending market next year is likely to intensify competition in this area yet further." This proved very perceptive in the light of subsequent events, and in 1991, giving evidence to the House of Commons Treasury and Civil Service Select Committee, Leigh Pemberton suggested that the sharp rise in house prices in the late 1980s had had major effects:

> Feeling richer, individuals saved less, and spent more. There was of course, always an element of illusion in this feeling of greater wealth: house owners benefited in a durable sense only if they were prepared to move downmarket or leave less to their heirs. But the impression, however false – of rising wealth – must have been a potent force encouraging higher levels of borrowing. (1991)

The Bank of England (1991a) later expanded on the Governor's oral evidence, noting that although an increase in house prices may create a false impression of wealth,

> the apparent increase in housing wealth in the late 1980s may have led people to feel that they needed to save less and could safely borrow more . . . The house price boom of the late 1980s considerably increased the wealth of the personal sector. In conjunction with strong real income

growth, this seems to have triggered a consumption boom financed by borrowing, and the personal sector moved into financial deficit. The magnitude of this deficit (£13bn in 1988) was unprecedented (1991a: 85).

This view has been reiterated by many well known economic commentators and effectively became the conventional wisdom in the late 1980s and early 1990s. The house price boom led to rising wealth, rising consumer spending, greater equity extraction and falling saving ratios. Chris Huhne (1992), then City and Business Editor of the *Independent*, stated:

> In the early 1980s, Britain was mired in its worst slump since the 1930s. Then came the housing boom. No one predicted its extraordinary force. Yet, by the end of the decade, the spiral of credit, house prices and personal wealth had transformed the problems of the economy. Instead of too little spending, the optimism engendered by rising wealth meant that there was suddenly far too much . . . Not surprisingly, home owners were prepared to spend some of those gains on consumer goods and services. In part they borrowed against the security of their homes. But . . . the rise in house values encouraged other forms of borrowing – consumer credit for example – and must have made many home owners happier about running down their cash savings.

Similar views were expressed by Hamish McRae (1992b) in the *Independent*: "The scale of the 1988–89 consumer boom was underestimated because forecasters did not fully appreciate the link between the housing market and people's willingness to spend. People were prepared to run down their savings partly because they felt that they had made such a large profit on their houses."

Put simply, it was argued that rising house price increased housing wealth, which made owners feel wealthier, allowed them to borrow against their equity or cut back savings, and thus fuelled consumption and economic growth. In addition, the increased level of housing transactions in the late 1980s generated consumer spending and growth.

The Home Ownership Market as a Cause of the Recession?

The slump brought all this to an abrupt halt. Sales volume slumped, equity extraction fell dramatically in real terms, consumer spending fell sharply and in the early 1990s the argument was reversed to help explain the recession. Falling house prices in southern Britain, high levels of mortgage debt, negative housing equity and rebuilding of household savings were, it was argued, leading to a reduction of consumer spending and economic slowdown. Will Hutton (1991), then Economics Editor of the *Guardian*, stated that:

> Over the next 6–9 months, the shape of the recovery will depend almost entirely on the behaviour of the British consumer and in turn on the housing

market. The housing market is important because with 70% of households owner occupied and 44% holding mortgages, the trend in prices and turn-over has a vital import on both the feel-good factor and incomes that affects consumption . . . The great British housing boom in 1987 and 1988 was the single most important factor then overheating the economy . . . Now we have a great British housing bust, holding back consumption as much as it was once stimulated.

Other commentators echoed this view, pointing to the rise in mortgage debt. Thus Hamish McRae (1992a), Economics Editor of the *Independent* stated:

The recovery is in its early stages, but there are several reasons why it will differ from the previous ones . . . There is the hangover of debts accumu-lated during the decade, the burden of which is increased because the main assets against which they are secured (properties) have fallen in price. Many who took out large mortgages between 1987 and 1989 cannot now move, because the value of their home is less than the mortgage. Faced with large debts, most people save to try to recover their position. The result is that the saving ratio which fell steadily through the Eighties, is climbing again. In 1980 people saved about 13% of their after-tax income. This fell to 5% in 1988, but has now recovered to more than 10%.

Bob Pannell (1992: 8) of the Council of Mortgage Lenders also suggested that: "The weakness of house prices and low level of housing market activity may well have caused the personal sector to reassess the realisability of its equity tied up in home ownership and concluded that for the time being it needs to have a larger stock of liquid financial assets than in the past." Foley (1991) argued in the *Lloyds Bank Bulletin* that: "Despite the weak state of the house market, fears are being expressed about the effects of a recovery, since it is now widely recognised that there are important links between this market and the wider economy". *The Economist* (1992b) added that: "The real worry about the housing slump in the south is its effect on the wider economy. It is not merely hurting mortgage lenders. It is delaying economic recovery by making people feel less wealthy and so more inclined to save. And lower turnover in the housing market depresses sales of the furniture and carpets people buy when they move home." Finally, John Major, Chancellor of the Exchequer, said in the Government's 1990 Autumn Financial Statement that: "The weak housing market has probably contributed to the moderation of consumer spending, just as the buoyancy of the housing market in the previous two years contributed to its earlier strength" (quoted in Costello, 1991).

Once again, the housing market was centrally implicated in changes in the wider economy. When it boomed, the economy boomed; when it slumped so did the economy. It would not be too much of an exaggeration to suggest that the home ownership market was almost seen as the key to the economy. How valid is this interpretation?

Home Ownership and the Wider Economy: An Assessment

The aggregate empirical evidence is persuasive. Home ownership levels rose from 55% in 1980 to 67% in 1990 and national average house prices doubled between 1983 and 1989. During the same period the value of physical assets (predominantly dwellings) owned by the personal sector doubled from £553 billion to £1,137 billion, the share of net personal wealth accounted for by dwellings rose from 44% to 52% and total net personal wealth almost doubled from £1,043 billion to £1,939 billion (CSO, 1991). Financial liabilities also doubled from £178 billion in 1984 to £343 billion in 1988, and house purchase loans as a proportion of financial liabilities rose from 61% in 1984 (£109 billion) to 70.5% (£242 billion). Although the rise in liabilities was more than matched by the rising value of assets, the ratio of debt to disposable personal income rose sharply during the 1980s, rising from 57% in 1980 to 115% in 1990. Simultaneously, the personal sector saving ratio (which measures personal sector savings as a percentage of personal disposable income) fell from a peak of 13% in 1981 to a low of 4% in 1988.

Real personal disposable income rose substantially from 1983, to a peak of 5.9% in 1988, as did consumer spending. Equity withdrawal rose rapidly in the 1980s as Chapter 6 showed (Figure 6.2) and accounted for almost 7% of consumer spending in 1988. Net mortgage lending also rose dramatically, doubling between 1980 and 1982, and doubling again by 1986–87. It reached a peak of £40 billion in 1988: almost six times the figure of £7.3 billion in 1980. It is clear from these figures that the 1980s were marked by a rapid rise in house prices, mortgage debt, housing wealth and consumer spending and falling saving ratios. But the boom was unsustainable, and the slump in the home ownership market since late 1988 has been accompanied by a number of other changes in personal sector behaviour. Bob Pannell of the Council for Mortgage Lenders (1992) suggests that "The sharp tightening of monetary policy in 1988–89 prompted a significant correction by the personal sector. With around half of net wealth accounted for by home ownership and lending for house purchase making up about two thirds of personal sector borrowing . . . the housing and mortgage markets have borne the brunt of this adjustment."

There is convincing evidence for this. The personal saving ratio rose sharply from a low of 4.1% in the first quarter of 1988 to 10.9% in the third quarter of 1991. Consumer spending fell in real terms in 1991, and despite the fall in interest rates, personal sector borrowing has fallen sharply. Net new lending for house purchase has fallen since 1988 and the amount outstanding on consumer credit agreements has also fallen. The late 1980s and early 1990s have been characterised by falling sales, house prices and housing wealth, high levels of mortgage debt, stagnant consumer spending and a recovery in the savings ratio as households cut borrowing and rebuilt their savings. The level of real equity withdrawal fell from a peak of almost £20 billion in 1989 to −£2.5 billion in

1993. Repossessions peaked at 75,000 per annum, and 8.7% of households with mortgages were two months or more in arrears in 1991 (Wilcox, 1995).

It could be argued, however, that the figures given above represent merely correlations rather than causal relationships and a number of econometricians have undertaken analyses to try to show these relationships more rigorously. Lee and Robinson (1990) began by suggesting that houses can be thought of as investment goods which provide both a financial return and a flow of housing services (use value). They pointed to the massive increase in equity extraction in the late 1980s and suggested that, as the number of last-time sellers is unlikely to vary much from year to year, moving home owners are likely to be an important source of variation. They also showed that the greater the number of movers, and the higher the level of turnover, the greater potential equity extraction, even when all sellers behave "responsibly" and extract less equity than the profit on their transaction. Consequently, a rise in mortgage borrowing to finance greater housing market turnover and equity extraction is likely to be linked to a fall in the savings ratio.

Lee and Robinson argue that it is not surprising that an increase in wealth should increase the demand for housing as supply is largely fixed in the short term and, as housing is an asset market, a rise in price does not always deter the would-be purchaser who may be attracted by the possibility of making a holding gain. But they suggest that the special feature of housing is that individuals can fairly easily borrow against enhanced asset values, thus housing booms have more effect on real spending than stock market booms. A regression equation was able to statistically explain 62% of the variation in new borrowing by changes in house prices, turnover and mortgage interest rate. Carruth and Henley (1990) attempted to estimate the effect of activity in the housing market on aggregate consumer spending and indirectly on personal saving behaviour. They argue that most macroeconomic forecasters failed to predict the late 1980s boom in spending because they did not incorporate housing wealth and equity extraction in their equations. They argue that the level of housing market activity and associated equity extraction was very important in explaining consumer spending in the late 1980s as "the housing market operates in such a way as to allow households additional freedom in their choice of income level. During periods of housing market boom, households may adjust their income upwards by moving house and withdrawing equity . . . for spending purposes" (1990: 30).

When this is done, Carruth and Henley's results suggested "the potential for a huge growth in HEW fuelling the spending boom" over the period 1986–89. They show that between 1982 and 1989 the proportion of houses sold relative to the total owner occupied housing stock rose from 12% to 15%. In 1988, the total value of traded housing equity was £100 billion, and they estimate households had the ability to withdraw up to 12% of this traded equity, a total of £12 billion a year. This may have added up to 4% of consumer spending or £8 billion in 1988, when it experienced remarkable growth of 7%. They speculated that "the dramatic termination of the housing boom, in response to a sharp tightening of monetary policy, combined with rising inflation should have a considerable dampening effect on consumer spending" (1990: 33).

Housing Wealth and Consumer Spending: The Contrary View

Not surprisingly, the argument outlined above is challenged by the Council of Mortgage Lenders who, like the Building Societies Association before them, reject any view which directly links mortgage finance or house-price inflation to economic problems. Thus, Costello and Coles (1991: 15) argue that:

> It has frequently been alleged that equity withdrawal by individual borrowing on mortgage for purposes other than house purchase was a major impetus behind the recent consumer boom. But a number of developments enhanced the personal sector's capacity to borrow for consumption reasons during the last decade . . . it is largely irrelevant that some borrowing was secured against privately-owned homes as the level of borrowing would probably have grown regardless of the form in which the assets were held . . . the housing market cannot be held to account for the recent consumer boom and present recession . . . house price inflation and consumer spending are largely a function of the same set of factors.

Similarly, Pannell (1992: 7), a CML economist, argues that "While the release of pent-up demand for mortgage finance and rising house prices were, in retrospect, a natural corollary of financial market deregulation, the effects were undoubtedly amplified by the . . . lax management of the macro economy in the mid to late 1980s."

The CML argument has some merit. It has frequently been assumed that, as (a) house prices and wealth have risen, (b) consumption spending is positively linked to wealth, and (c) consumer spending rose sharply in the late 1980s, then (d) the increase in consumer spending was a function of rising house prices. As Costello and Coles (1991: 14) point out: "The fact that these two variables move in tandem is not sufficient reason to conclude that the increase in housing wealth was the main factor explaining the rise in consumer spending during the late 1980s." The CML claim that rising consumer spending and rising house prices are both the result of the same set of factors. These are:

1. Lax monetary policy and low interest rates during the late 1980s.
2. Rapid growth in real incomes. The CML suggest that, as in the house price booms of the 1970s, rising house prices were accompanied by rising incomes and consumer spending.
3. Reductions in income tax during the late 1980s when the Government cut the basic rate of tax by 2% and greatly reduced higher rates of tax.
4. Financial deregulation, which increased the number and competitiveness of lenders and made borrowing easier.

Foley has also argued that the wealth effect of a house price boom can be overstated, and suggests that "Other factors may lie behind the sharp upturn in equity withdrawal in the 1980s . . . It seems more likely that it is de-regulation of the financial markets allied to the collapse of the building society cartel at the beginning of the decade, which is the main cause" (1991b: 2).

A rather similar view was taken by Paul Turnbull (1990), a City economic analyst. He accepted that "in a purely narrow and esoteric sense it is true that rapidly rising house prices did contribute to the generation of unsustainably rapid rates of growth in consumer spending, which in turn led to inflationary pressures . . . the high level of mortgage borrowing in this period tended to bolster consumer spending via the . . . familiar route of equity withdrawal." Turnbull states that many economists have consequently argued that, if only the upturn in house prices had been dampened down, then the overheating of the UK economy and subsequent inflation would have been less marked. But he states that this line of analysis "totally ignores the reasons why house prices were rising rapidly in the first place". Rapidly rising house prices were not generating an overheated economy, rather "an overheating economy was gener- ating rapidly rising house prices . . . the economy was overheating because of a loose monetary policy and unduly low interest rates" (1990: 8). According to Turnbull, rapid house price inflation was merely one symptom among many. Similarly, Cutler (1995: 260) argues that "the simultaneous boom in house prices and consumption in the second half of the 1980s should not be interpreted as evidence of a causal link between the two".

Turnbull and the CML are correct in pointing to the role of deregulation, lax monetary policy, the rapid growth of personal incomes and the higher rate tax giveaway to the better off (Hamnett, 1997a) in generating inflation, but there can be no doubt that the state of the home ownership market does have an impact on the wider economy. A high level of home sales generates housing-related spend- ing on home furnishings, consumer durables and DIY as well as a boom for the building industry and its materials suppliers. This spending will have multiplier effects throughout the economy as a whole. There can also be little doubt that a high level of sales combined with rapidly rising house prices allows home owners to extract much more equity from their homes and divert some of this to consumption spending. Rising prices and housing wealth also allow home owners to spend a higher proportion of their income on consumer spending.

Conversely, there is no doubt that a slump in the volume of transactions and falling house prices will have a generally depressive effect on the economy, particularly when many households have high levels of mortgage debt or neg- ative equity. Falling real wealth and greater debt will, other things being equal, tend to encourage households to rebuild their savings, cut debts and reduce consumption, although there is no academic research on this to date.

The Bank of England (1991a) pointed to the sharp fall in housing wealth and the growth rate in personal borrowing for consumption in 1990 and Huhne (1992) has referred to "debt deflation" which involves forced sale of assets, and as was pointed out in Chapter 4, the impact of negative equity on the finances of individual home owners has been considerable. What is disputed is the view that the home ownership market holds the key to economic recovery, and that all that is required to jump-start the economy is a surge in housing demand and sales. As Costello and Coles (1991) point out, the fallacy is that causation is seen to lead directly from the housing market to consumer spending and economic growth. While there is undoubtedly a link between the housing market and consumer spending, it is not the only one, nor necessarily the most important one. As Foley (1992: 3) pointed out: "A recovery in consumer spending does not appear to be dependent on a recovery in the housing market. Both are jointly dependent on a rebuilding of the personal sector's financial position as well as increased confidence about future economic prospects. Any housing market recovery will depend on potential buyers feeling confident enough to take on extra debt."

The problem with the view that the housing market is the key to consumer spending is that it ignores other, and arguably more important, causal links. In a nutshell it misinterprets an indirect secondary causal link as being the primary direct causal link. The home ownership market reinforces booms and slumps in the wider economy, but it does not directly cause them. For a more comprehensive analysis of determinants of consumer spending and economic growth it is necessary to look at changes in real income, and government fiscal and monetary policy. As Cutler (1995: 264) points out: "it is not clear how house price increases can cause consumption to rise, since in a general equilibrium framework, they are jointly determined . . . In other words, changes in house prices (and housing wealth) cannot be considered in isolation from the developments in the rest of the economy and, more often than not, they reflect – or act as a signal of – those developments." She also claims that, at an aggregate level, the effects of changes in wealth on consumption are debatable, insofar as home owners continue to demand the same quantity of housing services if prices rise. This seems to miss the point, however, in that the price of a house is not simply "the present discounted sum of the value of housing services derived from it". Houses are also stores of wealth and home owners can, and do, borrow, reduce savings or increase consumption against that wealth. Nor do owners have to trade down to release housing equity: they can simply trade equity for debt by taking out a larger mortgage, as Chapter 6 showed and as Cutler acknowledges. Figure 8.1 gives a very simplified (housing determined) model and a marginally more sophisticated model of the links between the home ownership market and consumer spending which points to the multiple determination of consumer spending.

The boom in the home ownership market in Britain in the mid 1980s was fuelled by several factors: first, financial deregulation in the early 1980s; second, a massive cut in higher rates of tax in the 1987 budget (abolition of marginal income tax rates of over 40%), which resulted in a large increase in disposable income for higher income groups; thirdly a general rise in disposable income; and fourthly the reduction in interest rates to a ten-year low in mid 1988. These

176

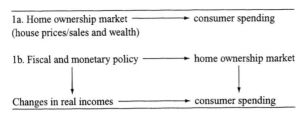

Figure 8.1 Two simplified models of home ownership and consumer spending

policies generated a housing boom of major proportions but they also generated a consumer boom of equal proportions. Not surprisingly, the increase in interest (and mortgage) rates from late 1988 to 1991 (from 10% to 15%), which was designed to cut consumer spending and reduce inflationary pressures, also had the effect of triggering a sharp contraction in the grossly inflated home ownership market.

A similar pattern can be discerned in previous booms. In the early 1970s a cut in interest rates led to the Barber boom and to the first house price boom (1971 to 1973), when national average house prices doubled. In response to inflationary pressures and high consumer spending the Government raised interest rates by five points in November 1973. The effect was the same as in the late 1980s. The overheated home ownership market collapsed almost overnight and prices and sales remained depressed for several years until 1976–77. By this time real incomes had risen sufficiently to bring the house price income ratio back to its historic level, house prices had fallen sharply in real terms and the overhang of unsold property had largely disappeared. The result was another housing boom which was itself curtailed by rising interest rates. In addition, both the early 1970s and the 1980s booms had a strong demographic component (Lee and Robinson, 1990; Ermisch, 1991). On this view, it can be argued that Britain's boom and bust economic policy has led to a boom and bust housing market, with marked secondary effects on consumer spending and the wider economy. To explain the booms and slumps in the wider economy largely in terms of booms and slumps in the home ownership market is to misinterpret cause and effect, though there can be no doubt that the housing market played a major contributory role.

There is no doubt that financial deregulation in the early 1980s led to a sizeable expansion of mortgage lending, which permitted house price inflation by virtue of the ease with which it was possible to get 95% mortgages and high mortgage to income ratios. House price inflation permitted owners to reduce their savings, extract equity and increase consumer spending. But to attribute these consequences solely to the housing market, and to ignore the effects of tax cuts, low interest rates and rising real incomes on consumer spending is to overlook the key determinants. The home ownership market reinforced the cyclical swings in the economy, but it has not been the sole cause of the boom and slump. As Costello and Coles (1991) point out: "Obviously the increase in house prices boosted housing wealth and this . . . exacerbated the inflationary effect of low interest rates. Events in the housing market ultimately took their cue from developments elsewhere in the economy including the existence of a lax monetary policy." It is to this policy and deregulation that we now turn.

The Role of Financial Deregulation in Promoting the 1980s Boom and Bust

Until the election of the Conservative Government in 1979 there was a formal Memorandum of Agreement between government and the building societies under which building societies gave preference to first-time buyers and raised funds for this rather than to finance equity extraction, but the new government allowed the agreement to lapse. Prior to this and other liberalisation measures, it was difficult for home owners to release any equity in their homes except by moving downmarket or taking out equity on a move.

The changes included the lifting in 1980 of the "corset" which had restricted the development of bank lending and borrowing, the entry of the banks into the mortgage market in 1981 so that they took 36% of net mortgage advances in 1982, and the 1986 Building Society Act which gave building societies access to wholesale funding and allowed them to lend for purposes other than house purchase as long as it is backed by housing collateral. As a result, says Muellbauer (1990c), "not only did financial liberalisation help to drive up house prices, it made housing wealth more spendable than ever before". He argues that land and housing prices are inherently unstable and that in the 1960s and 1970s the forces for instability were kept in check by a system of rationing of mortgage and other consumer credit. The abandonment of regulation and the rationing system has led to greater instability, an argument also made forcibly by Clapham (1996).

Many economists are agreed that financial deregulation and the availability of more generous mortgage credit permitted adjustment of consumer spending and savings preference in the 1980s whereas previously they had been much more restricted. Spencer and Scott (1990) argue that home owners engaged in forced savings in the 1970s because of credit rationing. The abolition of controls in the early 1980s had a major effect on housing investment and consumer spending as individuals were able to choose an optimum level of debt and consumption spending. The Bank of England (1991a: 85) reiterates this view:

> Borrowing by the personal sector against housing equity became easier during the 1980s as financial liberalisation fostered greater competition among banks and building societies . . . People were able to get loans representing a higher proportion of the value of their house or higher multiples of their income. There was a steady rise in persons' capital gearing (the value of debt for house purchase as a proportion of the owner occupied stock) through the early and middle 1980s.

Dicks (1987: 226) a Bank of England economist suggests:

One of the most important factors behind movements in the personal sector's balance sheet during the last decade has been change within the financial system, which has led to a weakening of the constraints which previously restricted households' choice. The sharp rise in both personal sector borrowing and liquid assets in recent years coincided with a period of rapid structural change and innovation in financial markets following a number of measures aimed at liberalising the market and encouraging competition. As a result mortgage queues have almost disappeared and, instead, interest rates are allowed to clear the market for funds. The increased competition led societies to lend higher multiples of income and to lower deposit requirements. Typically, multiples were around 2.5 during the 1970s, but have now risen to 3.5.

John Eatwell (1992) argued somewhat rhetorically that "The lethal cocktail of a heavily subsidised housing market and the deregulation of housing finance [were] major components of Nigel Lawson's debt-driven consumer boom. The resulting house price spiral encouraged and sustained the boom in consumer borrowing. Many people were driving around in their new extension."

But, as Miles (1992b: 100) points out:

households are now in a better position to make decisions on house purchase and on the size of their mortgages as part of a wider portfolio decision, rather than something solely to do with picking a place to live. The repercussions of this are great. Once the level of mortgage gearing on a house becomes a choice variable . . ..the options for lifetime consumption and savings look very different.

This is true but it has the effect, at least in the short term, of increasing "the spendability of illiquid assets, especially housing" (Muellbauer, 1990c). The optimistic scenario is that the shock of financial deregulation to the housing system was a one-off which will not be repeated and that households can make decisions over the balance between equity accumulation, debt and consumption which will not lead to an increase in instability. Both Clapham (1996) and Muellbauer take a more pessimistic view, however, arguing that the changes to the system caused by deregulation have led to a permanent higher level of instability. A new boom will lead to a rapid upsurge of new lending, rapidly rising prices and the prospect of subsequent slump, as lenders compete to retain or expand market share and households take out mortgages which they may not be able to repay if recession hits. If this view is correct, the 1980s boom and the 1990s slump may be the pattern of things to come rather than a one-off adjustment.

Miles (1992a: 65) points out that:

Housing wealth has become relatively fungible; it can be transformed into other commodities more easily than other forms of wealth important to the personal sector.... The fact that prior to the 1980s it was relatively hard for households to extract housing wealth ... means that the stock of net housing wealth held may have been ... significantly different from its equilibrium value.

Whether such crises will be as severe is debatable, however. I doubt that mortgage lenders would permit lending to rise so rapidly and so easily in the future as they did in the past.

The Housing Market and the Economy: A Problem caused by the South East?

One of the most frequent comments in the debate on the relation between the housing market and the wider economy is that the late 1980s boom and the subsequent bust were very much a southeastern phenomenon and that it was the boom and slump in the South East which effectively generated problems for the economy as a whole. Peck and Tickell (1992) suggest that the inflated economy of the South East was driving economic policy as a whole both in the upturn and the subsequent bust. As they put it: "Just as the boom of 1988 was a boom for the South first and for Britain second, so the slump of 1990 was a slump for the South first and Britain second. In this way a crisis of regulation within the core region effectively triggered a national accumulation crisis" (1992: 359). There is no doubt that it was in the South East that house prices rose most rapidly and fell most sharply and where consumer spending rose most rapidly and the savings ratio fell most sharply in the late 1980s. The savings ratio fell in all regions in the 1980s but the fall was much steeper in the South East than elsewhere and increases in personal disposable incomes and consumer expenditure were more marked (Hamnett, 1997a). To this extent, inflationary problems in the South East may well have triggered a national recession. It is well known that house prices rise in line with incomes in the medium to long term, but in the mid 1980s there was growing concern that rising house prices in the South East were leading to rising earnings as labour shortages grew, driven in part by the high price of houses and the difficulty of workers moving from other regions (Penycate, 1986; Hamnett, 1992b).

The existence of this phenomenon was analysed by Bover, Muellbauer and Murphy (1989), who showed that high house prices operated to crowd out manual workers from the South East, thereby adding to wage pressures in the region (Muellbauer and Murphy, 1991). They also argued that there are direct effects from house prices and mortgage costs on wage demands. In sum, the house price boom in the South East had impacts on the economy as a whole and arguably led indirectly to the subsequent recession as Nigel Lawson, the Chancellor of the Exchequer, raised interest rates sharply in 1988 and 1989 to choke off the inflationary boom which was driven by the rise in incomes, equity extraction and consumer spending in the South East.

One effect of the Conservative abolition of higher rates of income taxation in 1987 was to increase post-tax incomes in the South East, where high incomes are particularly concentrated (Hamnett, 1997a). Carruth and Henley (1993) specifically examined the regional dimensions of housing wealth, incomes and consumer spending in the 1980s. They found that regional disparities in the housing market increased rapidly in the 1980s, with substantial real growth of housing wealth in southern Britain. The level of housing equity in 1989 varied sharply from an average of £12,500 per person in the South East to just £2,500 in Scotland, partly reflecting regional differences in rates of owner occupation, and they found statistical evidence that regional differences in housing wealth affected consumer spending and savings ratios. To the extent that this is correct, it would appear that, at least in part, the early 1990s recession can be attributed to the impact of financial liberalisation, lax monetary policy and rising incomes and equity extraction from an inflated housing market in the South East of Britain during the 1980s. This did not cause the bust, but it may have been a contributory mechanism.

At the time of writing, prices are rising rapidly once again in London and the South East, which may prefigure the onset of a wider diffusion of price inflation. Given the detrimental impact of negative equity, particularly in the South East, this is no bad thing. It may help restore confidence in the housing market in the short term, but if it develops into another fully fledged consumer spending boom, with consequent Bank of England intervention via interest rates which depresses the whole country, it would be detrimental. The previous consumer booms all started in the South East but their consequences were nationwide. But we cannot simply blame the housing market. The concentration of high incomes in the South East meant that the Conservative Government's higher-rate income tax cuts in 1987 had a marked effect in the region, as did the marked increases in pre-tax incomes at the top end of the income distribution. The housing market in the South East is, in many ways, a reflection of the economic and social characteristics of the region, its economy and its inhabitants. For this reason future house price booms and busts are likely to start in the South East, particularly in the very expensive areas of central and inner London, where the concentration of high-income households is greatest, where effective demand and purchasing power are greatest, and where supply is largely fixed. To this extent, the home ownership market in the South East can be seen as a problem for the national economy.

CHAPTER 9

The Future of the Home Ownership Market in Britain

From the late 1960s to the late 1980s, the home ownership market in Britain has experienced a series of cyclical booms, the first in the early 1970s, the second in the late 1970s, and the third from the mid to the late 1980s. Although prices fell in real terms during the downturns, and substantially so in the mid 1970s, the dominant expectation of British home owners was of more or less permanent house price inflation which would, at minimum, keep pace with rising prices. Houses were viewed as a good investment, possibly the best there was for most people. Until the 1990s this was broadly correct, though most people were unaware of how far prices had fallen in real terms during the downturn of the mid 1970s. But in the 1990s the losses experienced by many new buyers were cash losses, not just real ones. For them, the housing market was a way of rapidly losing money and even of losing their home if they were unable to maintain the mortgage payments.

Not surprisingly, this has severely eroded confidence in the housing market and thus the key question is what is going to happen to the housing market over the next few years. Were the 1970s and 1980s an abnormal inflationary period? Or was the early 1990s slump an abnormal blip, and will it soon be back to business as usual, with house price inflation rising rapidly as we go into the next boom? The answers to this question vary dramatically. Some people believe that things have changed radically since the 1980s. Rachel Kelly (1992: 34) stated in *The Times* that "No one believes that house prices will boom any more. The bonanza of the 1980s was a one-off. When prices finally start to rise, they will do steadily and in line with inflation. Gone are the days when you could make more money by listening to the experts and waiting for their predictions of house price rises to come true." Arguably her view reflects the time. The British housing market was locked in severe depression and most people's memories are short. When prices are rising it is difficult to imagine that they could fall, and when we are in the depths of a slump it is difficult to imagine that the market could pick up again. But other voices have recently expressed similar views. In September 1994, Martin van der Weyer wrote an article in the *Spectator* titled "The Age of the Last Time Seller". In it, he argued that "a powerful conjunction of factors, some psychological, some demographic, are conspiring to kill off the idea of property ownership as an escalator to prosperity. . . . The aftermath of the property binge of 1986–89 has left lenders, borrowers, buyers and sellers all with blinding financial hangovers which will last until the end of century and beyond." He concluded that "the property market is dead; long live the roof over our heads" (1994: 20).

In 1995 van der Weyer went on to argue that we are entering a new world in the British home ownership market, one where:

> there will be no more booms and busts in . . . house prices. There will be fluctuations but property values should do no more than track the rate of real earnings growth. . . . Our housing market will become as placid as that of Germany. . . . Since the new generation of British homeowners will not have bought in the expectation of making spectacular gains, and will not be selling for a long time, they will gradually shed the obsession, inherited from the previous generation, with bricks and mortar as the essential form of tangible wealth and as a fabulous opportunity, every few years, to get rich quick.

More generally, Bootle (1995) and Wood (1995) claim that the economic climate has changed fundamentally, that inflation is dead, booms are a thing of the past and that houses are places to live in once more, rather than investment vehicles for owner speculators. Indeed, Wood (1995: 13) suggests in a well-argued article that "The world we are living in now bears no relation to the inflation-ridden Seventies or to Margaret Thatcher's debt-driven boom of the Eighties. House prices that struggle to hold their value in nominal terms, and are slowly shrinking in constant money terms, seem a much more likely prospect than any resumption of the boom."

Will the optimists be proved right? Do the 1990s mark the end of an era of housing price inflation, speculation and trading up? Will the demographics of a diminishing cohort of potential first-time buyers and a growing cohort of last-time sellers permanently transform the structure of the housing market in Britain? I have major doubts. While Bootle may be correct that we have entered a new low-inflation era, this does not necessarily mean that we are entering a new era of house price stability. The future of the home ownership market in Britain depends on which is the stronger of two opposed sets of forces. On the one hand, there are forces for stability which include the overhang of negative equity, the long-term demographic downturn, the shift to low inflation and the erosion of confidence in the housing market. On the other hand, there are still a strong preference for ownership, the legacy of the 1970s and 1980s booms on inflationary expectations of older owners, some potential pent-up demand from first-time buyers deterred from buying in the early 1990s, the continuing growth of real incomes, and the fact that housing still remains an asset market subject to speculative bubbles and booms and busts. The following sections look at each of these in turn.

Forces for Stability

The Maturing of the Home Ownership Market

For most of the postwar period, home ownership grew rapidly from a relatively small base – 35% of households – as a result of new building, the sale of private rented property and, in the 1980s and early 1990s, the sale of council housing, to its current level of 68%. Over this period, the number of home owners rose from 4 million to 16 million. This expansion is now largely over and the proportion of home owners has begun to stabilise as it has done in Australia, Canada, New Zealand and the USA, where the ownership level is around 65–70%. In Australia, home ownership has even fallen back slightly from its level of ten years ago. The high rate of council house sales during the 1980s which added 1.5 million home owners and raised the home ownership level by 5–6 percentage points is unlikely to be repeated, not least because many of the more attractive homes have been sold. Forrest and Leather (1995) point to the importance of "Right to Buy" in increasing the level of ownership among middle-aged and older households during the 1980s, and they point out that the next ten years will see a further increase in the number of households in this group. Nonetheless, the number of sales to sitting tenants has fallen from a peak of 190,000 in 1988–89 to 61,000 in 1993, and it is suggested that sales are unlikely to exceed 100,000 a year in the next few years. This would still add one million new owner occupied households over the next ten years, however, if sufficient sitting tenants decide that it is worth their while, or that of their children, to buy. Given the discounts, buying a good council house in an attractive area can be a good investment, though some sitting tenant buyers have found their purchases are millstones around their necks (Forrest and Murie, 1988). So too, the transfer of previously private rented dwellings into owner occupation is now largely complete (Hamnett and Randolph, 1987) and the private rented sector is now a small residual sector. Home ownership in Britain has largely ended a period of rapid expansion and is now on a plateau where most people who can afford to buy have bought. As Chapter 3 showed, home ownership levels have reached saturation point in the professional and managerial classes and expansion of home ownership into the skilled manual working class is now largely complete. The long-term potential to draw many more existing households into the tenure must be relatively limited. There are only a few countries where the home ownership rate surpasses that of Britain, and it is reasonably safe to assume that 70% of households may be the effective limit of home ownership.

Demographic Change

The growth in home ownership and the rise in house prices in the 1970s and the 1980s were fuelled, in part, by the growth in the number of first-time buyers entering the market. There was a sharp increase in the number of 20–29-year-olds from the late 1960s onwards as the postwar baby boom entered the housing market for the first time (Dicks, 1988; Breedon and Joyce, 1992; Cutler, 1995). In their econometric model of the UK housing market Milne and Clark (1990) found that "The long-run equilibrium is one in which real house prices depend on real income, inflation and the user cost (of housing) and the proportion of 25–29 year olds in the adult population". They added that "Changes in the latter have a marked impact on house prices with an increase in this proportion of 1 per cent . . . raising real house prices by 9 per cent. This proportion rose from 11 to 14 per cent from 1980 to 1989, accounting for a 30 per cent rise in real house prices during the 1980s. By comparison a 1 per cent rise in real incomes raises house prices by 1.2 per cent in the long-run" (1990: 36).

Not surprisingly, they stated that:

a key factor behind the extraordinary rise in house prices during the 1980s has been the baby boom of the 1950s and 1960s, with house prices being pushed up by as much as 20 per cent by the bulge of those born during the baby boom reaching the first-time buyer age group. *We believe that this premium will be unwound during the mid 1990s as the number in the first time buyer age group falls.* Capital gains on the scale made in the 1980s will not be repeated in the 1990s. (1990: 34, emphases added)

The importance of the size of the prime first-time buyer age group has been noted by other researchers. Lee and Robinson (1990: 45) found that "variations in demographic demand correlate fairly well with house price movements. The housing price explosion of 1972–3 occurred when those born in the post war baby boom reached house buying age. The recent one started when those born in the mid-1960s baby boom entered the housing market." They suggest that the imbalances are small, around 20,000 units at most compared with an annual flow of new buyers of 400,000–500,000, but, like Milne and Clarke they suggest that this imbalance of 5% can generate movement in the ratio of house prices to earnings of 20–30% as housing demand is highly inelastic and that the three house price booms of 1972–73, 1978–79 and 1985–89 have all been linked with increases in demographic demand.

Holmans (1995) shows that the annual number of births rose from 700,000 in 1941 to a peak of just over a million in 1947, falling to 800,000 in 1950–55, rising again to a peak of 1 million in 1964, and falling to a low of 657,000 in 1977 (Table 9.1). Such fluctuations will have an impact on potential housing demand twenty to twenty-five years later, and Holmans shows that the number of first-time buyers, excluding public sector sitting tenants, peaked at 500,000 a year in 1971 and 1972, falling to 350,000 in 1974, rising to 500,000 in the mid to late 1980s and falling back to around 380,000 in 1990s (Table 9.2). He argues that in the six years 1983–88 the total of first-time buyer purchases exceeded the 1974–80 average by almost 500,000, and almost half of this total can be

Table 9.1 Births in the UK in selected years (000)

1930–40 (Annual average)	725
1941	696
1944	878
1947	1,025
1950	818
1955	789
1964	1,015
1968	947
1974	737
1977	657

Source: Holmans (1995)

Table 9.2 Estimated number of house purchases for owner occupation in the UK, 1971–93 (000)

	Moving owners	First-time buyers (excluding PSSTs)	Public sector sitting tenants	Total	FTBs %
1971	425	497	19	941	53
1972	478	498	61	1,037	48
1973	429	402	41	872	46
1974	343	350	5	698	50
1975	505	456	2	963	47
1976	511	433	4	948	46
1977	532	426	12	970	44
1978	575	407	30	1,052	39
1979	542	406	42	990	41
1980	496	379	90	965	39
1981	569	371	124	1,064	35
1982	658	371	235	1,264	29
1983	758	444	170	1,372	32
1984	810	499	128	1,437	35
1985	820	496	108	1,424	35
1986	894	473	105	1,472	32
1987	1,016	481	122	1,619	30
1988	1,159	545	183	1,887	29
1993	566	382	61	1,009	38

Source: Holmans (1995)

explained as a result of population growth. About a quarter of the increase is a result of postponement of purchases from the early 1980s when the housing market was depressed and the bringing forward of purchases from the end of the 1980s in an attempt to "beat the boom". As a consequence of the earlier low birth rate and forward shifts of purchases the number of young first-time buyers in the early 1990s was very low. This was intensified by what may be deferral of

Table 9.3   The changing size of the 23–25 year age groups for selected years, UK

1974	2.415 m
1976–78	2.322 m (the low point of the 1970s)
1988	2.908 m (the peak)
1993	2.687 m
2001	2.060 m (the predicted low point)

Source: Holmans (1995)

purchases and independent living or a temporary shift to renting. As to the future, Holmans shows (Table 9.3) that, as a consequence of the change in birth rates, the size of the 23–25-year-old potential first-time buyer age group falls from a peak of 2.9 million in 1998 to a low point of 2.06 million in the year 2001: a fall of almost a third. Holmans suggests this will reduce the number of first-time buyers by around 50,000 a year by the late 1990s.

Ermisch (1991) examined the impact of demographic change on the home ownership market and pointed to the impact of the baby boom generation on the demand for housing. The rate of new household formation increased in the early 1970s as the original postwar baby boomers entered the housing market for the first time, but it rose even more steeply from 1982 to 1988 and, as Ermisch comments, "all else being equal, the maturing of the baby boom generations would be expected to increase house price inflation and housing construction activity" (1991: 233). But the number of first-time buyers rose far more rapidly during the 1980s than the number of new households, and Ermisch notes that other factors were at work which included "council house sales, particularly during the first half of the 1980s, the reduction in mortgage rationing and the inducement to become a home owner provided by a high expected rate of house price inflation during the 1983–88 house price boom" (1991: 233). Ermisch pointed out that in 1989 Britain was at the peak of new household formation from changes in the population age distribution and births in the late 1960s and he notes that "the maturing of the baby bust generations produces a steady decline in net annual household formation, from about 160,000 per annum in 1989 to about 40,000 per annum just after the turn of the century. This development increasingly puts downwards pressure on house prices and construction activity" (1991: 234).

Looking forwards, Ermisch notes that there is some recovery in household formation in the first decade of the next century, but it only reaches about half the level of the late 1980s, and during the second and third decades of the century net household formation could decline further to historically low levels. Thus, Ermisch concludes that "It appears that the 1980s are likely to have been the last of the demographically inspired housing booms." If Ermisch is correct, it would appear that the sudden demographic shocks to the home ownership market in the 1970s and 1980s are now over and it should prove more stable in future, though relatively depressed for the next few years. But a key issue will be the extent to which potential first-time buyers in their twenties who have deferred purchase in the early 1990s may enter the market if it appears that the slump is finally over. The other key issue will be the rate of new household formation. In recent decades, there has been a fall in average household size and strong growth in the number and proportion of one- and two-person households.

The latest household forecasts from the Department of the Environment (1996) suggest that England will need to build houses for an additional 4.4 million new households over the twenty-five years 1991–2016: an average of 220,000 new households a year, and the DoE is forcing planning authorities to provide for a much larger number of new dwellings in their structure/local plans. The Green Paper "Household Growth: Where Shall we Live?" indicated that 46% of the projected growth (just over 2 million) will occur as a result of an increasing population, of which approximately 500,000 is due to migration to England from other parts of the UK and from abroad, and some to people living longer. Another 33% (1.45 million) is the result of behavioural changes such as divorce and separation and just 21% (924,000) is due to changing age distributions, with more people entering the age groups which form households. Some 80% of the projected growth is of one-person households, of whom single, never married, men account for 1.4 million. These figures are strongly at variance with Ermisch's projections, which raises major questions as to whether new households really will form at the rate suggested by the DoE over the next twenty-five years. If they do, the underlying demographic pressure on the housing market could continue to be strong even though the population in the 20–29 age group is growing slowly. But as Bramley (1996) has shown, there is a circularity in the relationship beween the household projections and economic and housing market influences. The supply of housing influences the demand for it, as well as vice versa. As Downs (1980: 3) notes: "The tremendous acceleration of household formation in the United States during the 1970s resulted in part from the availability of housing at low cost (in real terms); it was not purely a demographic factor to which housing markets 'had to' accommodate themselves". In other words, if housing is cheap in real terms it may lead to an increase in the number of households to fill them. As the Green Paper itself acknowledges, the household projections are "essentially mechanical, based on a range of assumptions about social and demographic trends which may or may not materialise". The only thing which we can be certain about is the size of the key age groups in twenty years' time, as they have already been born. And, as Holmans (1995) has shown, the size of the key 23–25 age group will keep falling for the next few years as the "baby bust" of the mid 1970s reach their twenties and enter the housing market. The number of twenty-somethings will only begin to increase once again in the early years of the next century, as a consequence of the rise in birth rates in the late 1970s and early 1980s. This may be when house price inflation pressures will strongly reassert themselves.

A New Era of Low Inflation

The 1970s heralded an era of extremely high inflation with inflation reaching a peak of 24% in 1974. Inflation rose again in the early 1980s to 15% but it has fallen dramatically during the 1990s. Arguably, house prices rose sharply during the 1970s and 1980s partly in response to inflation and may stabilise now that inflation is lower and the inflation premium in house prices has disappeared. As Cutler (1995: 260) points out: "A more stable macro-economic environment in the 1990s – with lower and less volatile general price inflation – is likely to reduce the demand for housing as a hedge against inflation."

This argument was foreshadowed fifteen years earlier by Anthony Downs (1980: 3), who argued that one of the main effects of inflation is to magnify the benefits of investing in housing, as a small down payment magnifies return on owners' equity and reduces the carrying cost of housing, which, in turn, discourages consumers from saving out of current income. "They have simply moved their savings under their own roof . . . by regarding increases in their home equities as savings".

Cutler also notes that the nominal house price falls in the early 1990s may have raised the perceived risk of investing in housing relative to other assets, and the reduction in the tax advantages of investing in housing may reduce its attractiveness. In addition, she argues that lower general inflation should lead to lower and less variable interest rates which would reduce the problem of the "front-end loading" of mortgage repayment costs to the early years. This should increase household ability to service debt. The lower expected returns to owning a home financed by a mortgage should also reduce the incentive to households to increase their mortgage debt in order to maximise their returns.

The Erosion of Confidence in the Stability of the Home Ownership Market

Prior to the early 1990s the phrases "as safe as houses" or "as safe as bricks and mortar" were not subject to serious question despite the substantial falls in real prices in the mid 1970s. The fall in nominal house prices and the rise of negative equity and repossessions have undermined the near certainty that house prices would rise for a generation of first-time buyers, and their expectations of home ownership as a source of long-term gains have been shattered. The short-term preference for home ownership has fallen significantly among 20–29-year-olds, though whether this will prove to be permanent remains unclear. As Clapham (1996) rightly argues, the reduction in other housing tenure opportunities as a result of the "Right to Buy" and the long-term decline of private renting means that most people today have little option but to buy, like it or not, and in the medium term there may be little reduction in home ownership rates, though it remains doubtful how many of the 400,000 who have suffered repossession will, or can, ever re-enter the home ownership market, not least given their credit-ratings with mortgage lenders. More generally, as Cutler, van der Weyer and others suggest, the long-term impact of the slump may be to turn home owners' attention away from an obsession with asset values to viewing housing as providing a roof over your head.

The One-Off Impact of the Liberalisation of Housing Finance in the Early 1980s

It has been suggested that the liberalisation of the housing finance market in the early 1980s is now behind us, and that we are now unlikely to experience another shock of this nature, not least because the financial institutions have introduced tougher lending criteria in the 1990s in an attempt to prevent a re-occurrence of the boom and bust of the 1980s. From this perspective, the re-adjustment of household debt, savings and consumption patterns was a one-off. Chrystal (1992: 39) suggests, for instance, that "the combination of factors which triggered the (1980s) bubble is unlikely to recur this century". But, as Miles (1992a), Clapham (1996), Muellbauer (1990c) and others argue, the greater availability of mortgage finance, and the competition between lenders for market share are such that a return to boom conditions could easily induce a surge in new lending and house prices. In this view, permanent instability has been introduced into the system and it can easily break out once again.

Forces for Instability

Over the past thirty years at least, house prices and incomes have tended to move cyclically. First real incomes rise while house price inflation remains low, then as incomes feed into the market, house prices rise quite steeply to a peak. Then, as the house price income ratio peaks, house prices mark time or fall in real terms while incomes begin to rise above house prices again. At present, we are at the bottom of the cycle. Houses are currently very cheap relative to incomes, but they have begun to rise sharply again in London and the South East as they did in the three booms of the 1970s and 1980s for the reason that incomes are highest in London and rise most rapidly there. If the past is any guide to the future, then we may expect rising prices in the rest of the country in the next few years. But how rapidly prices will rise depends both on incomes and on demography as, expectations aside, these are arguably the two most important forces which drive house prices. As Chapter 2 noted, it is ability to borrow and ability to pay (i.e. income) which influences how much people are able to pay for houses. Expectations determine how much they are willing to pay. If the expectation is that prices will continue to rise, then most people are willing to pay more up to the limit of their ability to borrow. But if the expecta-tion is that prices will remain broadly stable, most people will not be willing to pay inflated prices. In this sense, expectations are quite fundamental, and Milne and Clark (1990) claim that "The short-run dynamics of house price movements are determined by expectations of capital gain". But expectations are also dependent on prices and prices are in turn dependent on the balance of demand and supply.

The supply of houses is broadly fixed. The private house building industry built between 141,000 and 147,000 houses per year from 1992 to 1996. In the boom year of 1988 it reached 200,000. This compares to sales of around 1.2 million a year at present and 2 million in the peak year of 1988. At best, therefore, the house building industry contributes between one-sixth and one-eighth of all sales. Of course, many of these sales are between existing home owners rather than to first-time buyers, and as Chapter 7 showed the number of first-time buyers was crucial in helping fuel market demand in both the early 1970s and the mid 1970s. The number of last-time sellers is relatively fixed, but the number of new potential first-time buyers depends on both the underlying growth rate of the key first-time buyer age cohort and on the rate of new household formation. Although the former is given by the number of births twenty to thirty years previously, the rate of new household formation is more variable, as is propensity for households to become home owners rather than renters or sharers. It is conceivable that the housing slump of the early 1990s acted to temporarily deter many potential first-time buyers from entering the market. There may, therefore, be a substantial pent-up potential demand if these potential buyers have not been permanently deterred (Holmans, 1997).

Expectations of future price movements are fundamental to the existence of booms and slumps in the home ownership market as in all asset markets. In a study of 1,500 properties in the Boston area that were sold more than once between 1978 and 1985, Case (1988) tried to analyse the factors which might have produced the overall inflation of almost 60%. He looked at a number of fundamentals which are likely to affect the demand for housing. These included population growth, employment and income, construction costs, interest rates, demographic change, lower taxes, fuel and utility costs. He used these factors to produce a structural model of the housing market which used pooled data on 11 US cities over 10 separate years from 1971 to 1980. The model was used to retrospectively "predict" the amount of house price inflation between 1980 and 1985 both for the US and for Boston. But the actual degree of price inflation in Boston was far greater than predicted.

Consequently, he put forward an explanation based on rational choice in an inflationary environment. This model, based on anticipated inflation, leads to inflationary "asset price bubbles", that is to say that people's expectations of asset price inflation will bring about high asset price inflation, and he argued that "the housing market is more likely than others to generate price bubbles", not least because information on yields is always imperfect and because real estate agents, who sell on fixed commission, have a vested interest in driving prices

192

higher in order to "test" the upper limit of the market. As Case notes: "If you overprice, you can always cut the asking price; if you underprice, and the house sells in 10 minutes, you can't change your mind" (1988: 45). Thus, he suggests that "in serving the best interests of their clients, agents are likely to generate perfectly rational expectations of future increases". He adds that "It is very doubtful that real estate agents can start an expectations spiral, but if market fundamentals begin to generate increases . . . the potential is there for perfectly rational buyers, sellers and agents to turn those increases into an expectational bubble" (1988: 46).

Case argued that inflationary spirals can continue as long as ability to pay, income, wealth and borrowing can support them. In a study of four American cities, two undergoing booms (Anaheim and San Francisco in California), a post-boom market in Boston, and a stable market (Milwaukee), Case and Shiller (1988) examined the attitudes and behaviour of some 900 home buyers in 1988. They suggest that housing investors are not aware of fundamentals. Instead, they "tend to interpret events in terms of hearsay, cliches, and casual observations". They also found that investment motives were high on buyers' incentives, and suggest that "buyers in booms expect still more appreciation of housing prices and are worried about being priced out of the housing market in the future". Thus, there is a strong role for expectations. They argued that in general the demographic or income triggers for booms are not observed directly by investors except through prices.

The issue of demographic change has already been discussed. Real incomes have continued to rise in the early 1990s at around 2.5% annually, and the national house price/earnings ratio is almost back at the levels of the early 1970s, mid 1970s and the early 1980s. This suggests that potential purchasing power is there to support another sharp rise in house prices, and if we accept Saunders's (1990) argument about the long-term relationship between incomes and house prices then house prices are bound to rise in the medium to long term. Will the inflationary booms of the 1970s and 1980s reoccur?

Clapham (1996) suggests that the Government's view of the role of housing in the 1980s boom is that it was caused by a series of mistakes that will not reoccur. Government failed to incorporate the links between housing and the wider economy into its economic models, buyers overextended themselves and took on too much debt, and lenders lent money too readily and with too few safeguards and are now paying for their mistakes. But Clapham argues that "far from being a one-off abberation, the problems which have occurred since the mid 1980s are . . . an inevitable consequence of the deregulation and the increasing role of the market in the British housing system" (1996: 637). His argument has three strands. The first concerns the behaviour of households, the second the behaviour of financial institutions, and the third focuses on the nature of the economic system which has been created, particularly the "so-called flexible labour market". He claims that the view that households will change their behaviour and cease to view home ownership as an investment is flawed, and suggests that while the last few years "may have changed people's attitudes towards the reasons for wanting to become home-owners . . . it is unlikely to change their wish to enter the sector" (1996: 637). Although owner occupation may no longer be seen as the path to quick riches that it may have seemed to some in the late 1980s, this will not result in any substantial fall in demand for the sector because of the lack of alternatives. Clapham argues that the most important fact is the cyclical nature of the home ownership market, which reinforces expectations and means that the system is inherently unstable. The degree of instability is likely to depend, says Clapham, on the the ability of households to bring forward or postpone purchases at different stages of the cycle.

The second strand of Clapham's argument concerns the key role of financial institutions. He argues that in the past the inherent instability in demand for owner occupation was moderated by regulation of mortgage lending by both government and the building societies and the existence of mortgage queues which stabilised and choked off excess demand in times of rising prices. This was replaced in the early 1980s with a system which is based on competition between lenders and "creates incentives to increase the availability of funds at these times". He points out that the proportion of total lending by the building societies fell from 80% in 1980 to 60% at the height of the boom in 1988 and to 51% in 1993. He argues that "the battle to keep out competitors and retain market share will lead to increased lending at a time of increasing prices" (1996: 639). The mortgage lending market is now more competitive than ever before, and the conversion of building societies into banks means that shareholder pressure to perform will be much stronger.

Clapham's third argument in support of an unstable housing system is the creation of the "flexible labour market", which has involved deregulation, the weakening of the power of trade unions, the abolition of wages councils and other things. The result has been an increase in job insecurity and the growth of self-employment and part-time work. But, says Clapham, home ownership is not easy to sustain in an insecure context, particularly in Britain where the Government has cut back the payment of state benefits for mortgage interest payments to those who have lost their jobs. This means that in the event of another recession, "many home-owners will lose their homes along with their jobs". Clapham concludes that "The result of the British approach to the creation of a competitive market economy and the privatisation and deregulation of housing is a mutually reinforcing and unstable system in which economic cycles will be heightened by the housing system and both will be characterised by large fluctuations in activity" (1996: 640).

There has been a marked shift in recent years away from secure permanent jobs to what is sometimes termed a more "flexible" labour market, which often means in practice a greater degree of job insecurity and an intensified risk of unemployment. This has implications for the ability to take out and repay a mortgage, as the mortgage lenders are just beginning to appreciate. As Ford (1995: 6) points out:

> For much of the postwar period, mortgage lenders have assumed that their borrowers would be continuously employed in secure jobs with rising incomes. Any employment mobility was likely to be voluntary, the risk of unemployment limited, and if it did occur, short term. Lending over a 25-year period was therefore appropriate and relatively risk free. In part this resulted from the fact that until the late 1970s, mortgagors were over-whelmingly drawn from professional, managerial and other higher white collar occupations, groups who were largely immune to cyclical unem-ployment. These certainties were matched by the security . . . derived from a high inflation economy where the value of the debt was quickly eroded whilst the value of the property rose, but also from the existence of the state safety net. This met some or all of the interest costs for mortgagors who . . . found themselves unemployed.

Ford notes that in recent years these certainties have lessened or disappeared. First, the expansion of owner occupation, partly through the "Right to Buy", has drawn in borrowers who face higher risks of unemployment. Second, the high-inflation economy has given way to a low-inflation economy, where debt is slow to erode, and changes in the state safety net for claimant mortgagors introduced in October 1995 have considerably reduced the state's willingness to pay mort-gage interest for unemployed borrowers. She also points out that the labour market has changed significantly, with a long-term increase in less secure part-time jobs and a greater likelihood of management job losses. She notes that one in four men between the ages of 16 and 64 are without a job and the number of people classified as economically inactive is rising. In addition to the 1.6 million men officially registered as unemployed, a further 2.2 million men of working age are outside the labour force. In sum, although full-time permanent jobs re-main the norm, there is "a clear growth in more precarious forms of employment", including part-time, temporary, short-duration and self-employment. Although these changes are concentrated in the rented sector they are percolating into the owner occupied sector and are likely to reduce people's confidence in their ability to take out a mortgage and meet long-term continuing repayments. The British Social Attitudes Surveys show that the proportion of people who agree that buying a home is too much of a risk for a couple without secure jobs rose from 59% in 1986 to 69% in 1991. Given that labour market insecurity is likely to become more rather than less pronounced over the next decade, this is a problem for home ownership. In their logistic regression model of the social distribution of mortgage arrears, Burrows and Ford (1997) found that lone par-enthood increased the probability of arrears of more than three months by a factor of 4.8 times, being employed part-time increased it by 2.9 times, being unable to work increased it by 5.2 times, and being unemployed by a remarkable 9 times.

Cutler (1995) argues that although lower general inflation may be expected to stabilise the market, it is unlikely to lead to the disappearance of booms and slumps given that housing still functions as an asset market. Indeed, she argues that whereas high inflation disguised real price falls in the 1970s and 1980s (in 40% of years since 1943), this is unlikely to happen in the future and nominal price falls may be more marked in a low-inflation environment, which may contribute to greater instability in the market. This may also serve to weaken people's housing expectations and may lead to less overinvestment for speculative purposes. Meen (1995) adds that the rise in gearing (mortgage debt to income) ratios and the absence of mortgage rationing which resulted from the liberalisation have meant that "the housing market and the economy in general have become more sensitive to changes in interest rates than in the 1970s" (1995: 408).

The depressed nature of the housing market through the early 1990s and the growth of negative equity and repossessions led to a general desire for a healthy bout of renewed house price inflation that would kickstart the market, reduce negative equity and reinject confidence in the long-term investment potential of home ownership. Measures of house price/earnings ratios (the standard proxy for affordability) show sharp falls since 1988, particularly in London and the South, and suggest that there is considerable potential for an upturn in housing prices. This appears to have begun in mid 1995, although it has been restricted to London and the South East at the time of writing. How far it will proceed remains open to debate and one press article at the time of writing was titled "The boom that never was" (*Independent*, 30 November 1997). It is doubtful, however, whether we will see the advent of the stable housing market predicted by van der Weyer and others. The lull before the storm may prove to be a more accurate assessment. Housing is now recognised to be an asset, and asset markets are prone to speculative bubbles, particularly when the finance is there to support asset buying. If house prices keep rising strongly in London and the South East during the late 1990s, we may well see the rapid diffusion of house price inflation across Britain once more by the turn of the century as first-time buyers once again feel the urgency to get on the ladder before it disappears out of financial reach.

The key question regarding the next house price boom is therefore not if, but when. Memories are notoriously short-term, and rising prices could propel many new buyers into the market once again. All we can hope for is that the lenders have learnt their lesson about lending on high price/income and advance/price multiples. A return to the need for 5–10% deposits of the 1960s and 1970s would be no bad thing, as this would prevent buyers over-extending and provide protection against the return of the negative equity nightmare of the early 1990s. The biggest factor which militates against another boom of 1970s or 1980s proportions is the decreasing number of 20–29-year-olds in the population until the early years of the next century. However, a combination of rising real incomes, an increase in the size of the key house buying age group, low mortgage interest rates and a continuing low house price/earnings ratio could collectively fuel a boom of considerable proportions in the early years of the next century.

References

Abelson, P. (1997) House and land prices in Sydney from 1931 to 1989, *Urban Studies*, 34, 9, 1381–1401

Agnew, J. (1981) Homeownership and the capitalist social order, in Dear, M. and Scott, A.J. (eds) *Urbanization and Planning in Capitalist Society*, London, Methuen

Anderson, M., Bechhofer, F. and Kendrick, S. (1990) Individual and Household Strategies: some empirical evidence from the social change and economic life initiative, unpublished paper

Atkinson, A.B. (1983) *The Economics of Inequality*, 2nd edn, Oxford, Clarendon Press

Atkinson, A.B. and Harrison, A.J. (1978) *Distribution of Personal Wealth in Britain*, Cambridge, Cambridge University Press

Atkinson, A.B, Gordon, J.P. and Harrison, A.J. (1989) Trends in the shares of top wealth-holders in Britain, 1923–1981, *Oxford Bulletin of Economics and Statistics*, 51, 3, 315–32

Badcock, B. (1989) Homeownership and the accumulation of real wealth, *Environment and Planning D, Society and Space*, 7, 1, 69–91

Badcock, B. (1994a) Urban and regional restructuring and spatial transfers of housing wealth, *Progress in Human Geography*, 18, 3, 279–97

Badcock, B. (1994b) "Snakes or Ladders?": the housing market and wealth distribution in Australia, *International Journal of Urban and Regional Research*, 609–27

Ball, M. (1982) Housing provision and the economic crisis, *Capital and Class*, 17, 66–77

Ball, M. (1983) *Housing Policy and Economic Power*, London, Methuen

Ball, M. (1986) Mortgage finance and owner occupation in Britain and West Germany, *Progress in Planning*, 26, 185–260

Ball, M. (1993) *Under One Roof: retail banking and the international mortgage finance revolution*, London, Harvester Wheatsheaf

Bank of England (1985) The housing finance market: recent growth in perspective, *Bank of England Quarterly Bulletin*, March, 25, 1

Bank of England (1991a) Recent sectoral financial behaviour, *Bank of England Quarterly Bulletin*, February, 31, 1

Bank of England (1991b) Recent developments in the UK housing market, *Bank of England Quarterly Bulletin*, August, 336–40

Bank of England (1992) Negative equity in the housing market, *Bank of England Quarterly Bulletin*, August, 32, 3, 266–9

Beaverstock, J., Leyshon, A., Rutherford, T., Thrift, N. and Williams, P. (1992) Moving houses: the geographical reorganisation of the estate agency industry in England and Wales in the 1980s, *Transactions Institute of British Geographers*, 17, 2, 166–82

Bellman, H. (1927) *The Building Society Movement*, London, Methuen

Bentham, G. (1986) Socio-tenurial polarisation in the United Kingdom, 1953–83: the income evidence, *Urban Studies*, 23, 1, 57–62

Blackmore, K. (1984) *An Econometric Model of the UK Housing Market: 1970–1982*, Discussion Paper No 13, London, Henley Centre for Forecasting

Boddy, M. (1980) *The Building Societies*, London, Macmillan

Boddy, M. (1989) Financial deregulation and UK housing finance: government–building society relations and the Building Societies Act 1986, *Housing Studies*, 4, 2, 92–104

Boleat, M. (1994) The 1985–1993 housing market in the United Kingdom: an overview, *Housing Policy Debate*, 5, 253–74

Bootle, R. (1995) *The Death of Inflation*, London, Nicholas Brealey

Bootle, R. (1997) House prices face the broom cupboard test, *The Times*, December

Bourdieu, P. (1984) *Distinction: a social critique of the judgement of taste*, London, Routledge

Bover, O., Muellbauer, J. and Murphy, A. (1989) Housing, wages and UK labour markets, *Oxford Bulletin of Economics and Statistics*, 51, 2, 97–136

Bowen, A. (1994) Housing and the macro-economy in the United Kingdom, *Housing Policy Debate*, 5, 241–52

Bramley, G. (1988) *Access to owner occupation*, Research Note, Association of District Councils, London

Bramley, G. (1993) The impact of land use planning and tax subsidies on the supply and price of housing in Britain, *Urban Studies*, 30, 1, 5–30

Bramley, G. (1994) An affordability crisis in British housing: dimensions, causes and policy impacts, *Housing Studies*, 9, 1, 103–24

Bramley, G. (1996) *Housing with Hindsight: household growth, housing need and housing development in the 1980s*, Council for the Protection of Rural England, December

Breedon, F.J. and Joyce, M.A. (1992) House prices, arrears and possessions: 3 equation model for UK, *Bank of England Quarterly Bulletin*, May, 32, 2, 173–9

Buckley, R. and Ermisch, J. (1982) Government policy and house prices in the United Kingdom: an econometric analysis, *Oxford Bulletin of Economics and Statistics*, 44, 4, 273–304

Buckley, R. and Ermisch, J. (1983) Theory and empiricism in the econometric modelling of house prices, *Urban Studies*, 20, 83–90

Building Societies Association (1983) *Housing Tenure*, London, BSA

Building Societies Association (1984) *Housing Finance into the 1990s*, London, BSA

Building Societies Association (1984) *Housing Finance*, London, BSA

Burchardt, T. and Hills, J. (1997) Mortgage payment protection: replacing state provision, *Housing Finance*, 33, 24–31

Burnett, J. (1986) *A Social History of Housing, 1815–1985*, London, Methuen

Bush, J. (1995) From subsidy to subsidence: a Tory approach to housing, *Independent*, 5 May

Callen, T.S. and Lomax, J.W. (1990) The development of the building society sector in the 1980s, *Bank of England Quarterly Bulletin*, November, 503–10

Carruth, A. and Henley, A. (1990) The housing market and consumer spending, *Fiscal Studies*, 27–38

Carruth, A. and Henley, A. (1993) Housing assets and consumer spending: a regional analysis, *Regional Studies*, 27, 7, 611–21

Case, K. (1986) The market for single family homes in the Boston area, *New England Economic Review*, May/June, 40–48

Case, K. (1992) The real estate cycle and the economy: consequences of the Massachusetts boom of 1984–87, *Urban Studies*, 29, 171–84

Case, K. and Shiller, R. (1988) The behaviour of home buyers in boom and post-boom markets, *New England Economic Review*, November/December, 29–45

Champion, A., Green, A. and Owen, D.W. (1988) House prices and local labour market performance, *Area*, 20, 3, 253–63

Checkoway, B. (1980) Large builders, federal housing programmes and post-war suburbanization, *International Journal of Urban and Regional Research*, 4, 21–45

Choko, M. (1995) Home owners: richer or not – is that the real question, in Forrest, R. and Murie, A. (eds) *Housing and Family Wealth: comparative international perspectives*, London, Routledge

Chrystal, K.A. (1992) The fall and rise of saving, *National Westminster Bank Quarterly Review*, February, 24–40

Cicutti, N. (1994) Estate agents are three a penny, *Independent*, 12 October

Cicutti, N. and Willcock, J. (1994) Nationwide gives up on estate agencies, *Independent*, 12 October

Clapham, D. (1996) Housing and the economy: broadening comparative housing research, *Urban Studies*, 33, 4/5, 631–47

Coles, A. (1992a) How mortgage possession rescue works. Mortgage rescue schemes: an overview, *Housing Review*, 41, 5, September–October, 85–6

Coles, A. (1992b) Causes and characteristics of mortgage arrears and possessions, *Housing Finance*, 13, February

Congdon, T. (1995) Unsafe as Tory houses: voters are blaming the government for falling house prices, *The Times*, 16 March

Congdon, T. and Turnbull, P. (1982) The coming boom in housing credit, London, Messell and Co., June

Coombes, M. and Raybould, S. (1991) Local trends in house price inflation, Joseph Rowntree Foundation, York, Housing Research Findings, 30

Cooper, C. (1976) The house as symbol of self, in Proshansky, H., Ittelson, W. and Rivlin, L. (eds) *Environmental Psychology*, 2nd edn, New York, Holt Rinehart and Winston

Cornford, J. and Dorling, D. (1995) Who has negative equity? How house price falls in Britain have hit different groups of home buyers, *Housing Studies*, 10, 151–78

Costello, J. (1991a) The housing market in the fourth quarter of 1990, *Housing Finance*, 9

Costello, J. (1991b) House prices and earnings, *Housing Finance*, 11, 8–16

Costello, J. and Coles, A. (1991) The housing market and the wider economy, *Housing Finance*, February, 14–19

Council of Mortgage Lenders (1989) *Housing Finance*, 4, October

Council of Mortgage Lenders (1992) *Housing Finance*, 13, February

Council of Mortgage Lenders, (1997) *Housing Finances*, 34, May

Counsell, G. (1992) Nightmare on Acacia Avenue, *Independent on Sunday*, 16 August

Couper, M. and Brindley, T. (1975) Housing classes and housing values, *Sociological Review*, 23, 563–76

Coyle, D. (1997a) The home ownership boom is pushing up unemployment, *Independent*, 25 July

Craig, P. (1986) The house that Jerry built? Building societies, the state and the politics of owner-occupation, *Housing Studies*, 1, 1, 87–108

Crompton, R. (1986) Consumption and class analysis, in Edgell, S., Hetherington, K. and Warde, A. (eds) *Consumption Matters*, Oxford, Blackwell

Crow, G. (1989) The use of the concept of strategy in recent sociological literature, *Sociology*, 23, 1–24

CSO (1991) *Economic Trends*, Annual Supplement, HMSO

Curtice, J. (1991) in Curtice, J. and Jowell, R. *British Social Attitudes Survey: the 1991 Report*, Aldershot, Gower

Cutler, J. (1995) The housing market and the economy, *Bank of England Quarterly Bulletin*, August, 35, 3, 260–69

Daunton, M. (1987) *A Property Owning Democracy*, London, Faber

Davies, R.B. and Pickles, A.R. (1991) An analysis of housing careers in Cardiff, *Environment and Planning A*, 23, 629–50

Davidson, J., Hendry, D., Srba, F. and Yeo, S. (1978) Econometric modelling of the aggregate time series relationship between consumers' expenditure and income in the UK, *Economic Journal*, 80, 661–92

Davis, E.P. and Saville, I.D. (1982) Mortgage lending and the housing market, *Bank of England Quarterly Bulletin*, September, 390–98

Davis, G. (1992) It could never happen here, *Independent*, 2 November

Department of the Environment (1977) *Housing Policy Review*, London, HMSO

Department of the Environment (1996) *Household Growth: where shall we live*, HMSO, November

Deverson and Lindsay (1975) *Voices from the Middle Classes*, London, Hutchinson

Dicks, M.J. (1987) The financial behaviour of the UK personal sector, *Bank of England Quarterly Bulletin*, May, 27, 2, 223–33

Dicks, M.J. (1988) The demographics of housing demand: household formations and the growth of owner occupation, *Bank of England Discussion Paper*, 32, July

Dicks, M.J. (1989) The housing market, *Bank of England Quarterly Bulletin*, 29, 1, February, 66–75

Dicks, M.J. (1990) A simple model of the Housing Market, *Bank of England Discussion Paper*, 49, May

Dieleman, F. (1992) Struggling with longitudinal data and modelling in analysis of residential mobility, *Environment and Planning A*, 24

Diggens, J.P. (1978) Barbarism and capitalism: the strange perspectives of Thorstein Veblen, *Marxist Perspective*, 1, 2, 138–55

Doling, J. (1993) British housing policy, 1984–1993, *Regional Studies*, 27, 6, 583–8

Doling, J., Karn, V. and Stafford, B. (1985) How far can privatisation go? Owner occupation and mortgage default, *National Westminster Bank Quarterly Review*, August, 42–52

Doling, J., Karn, V. and Stafford, B. (1986) The impact of unemployment on home ownership, *Housing Studies*, 1, 1, 49–59

Doling, J., Ford, J. and Stafford, B. (1988) *A Property Owning Democracy*, Aldershot, Gower

Doling, J., Ford, J. and Stafford, B. (1991) The changing face of home ownership: building societies and household investment strategies, *Policy and Politics*, 19, 2, 109–18

Donnison, D. and Ungerson, C. (1982) *Housing Policy*, London, Penguin

Doogan, K. (1996) Labour mobility and the changing housing market, *Urban Studies*, 33, 2, 199–222

Dorling, D. (1993) *The Spread of Negative Equity*, Housing Research Findings no 101, York, Joseph Rowntree Foundation

Dorling, D. and Cornford, J. (1995) Who has negative equity: how house price falls in Britain have hit different groups of home buyers, *Housing Studies*, 10, 2, 151–78

Downs, A. (1980) Too much capital for housing? *The Brooking Bulletin*, 17, 1, Summer, 1–5

Drayson, S.J. (1985) The housing finance market: recent growth in perspective, *Bank of England Quarterly Bulletin*, March, 80–91

Duncan, S. (1990) Do house prices rise that much? A dissenting view, *Housing Studies*, 5, 3, 195–208

Dunn, R., Forrest, R. and Murie, A. (1987) The geography of council house sales in England – 1979–1985, *Urban Studies*, 24, 47–59

Dupuis, A. (1992) Financial gains from owner occupation: the New Zealand case, 1970–88, *Housing Studies*, 7, 1, 27–44

Dupuis, A. and Thorns, D.C. (1996) Meanings of home for older owners, *Housing Studies*, 11, 4, 485–501

Dupuis, A. and Thorns, D.C. (1997) Regional variations in housing and wealth accummulation in New Zealand, *Urban Policy and Research*, 15, 3, 189–202

Earley, F. (1996) Leap-frogging in the UK housing market, *Housing Finance*, 32, 7–15

Earley, F. and Mulholland, F. (1995) Women and mortgages, *Housing Finance*, 25, 21–7

Eatwell, J. (1992) Britain doesn't need another housing boom, *The Observer*, 23 August

Economist (1988a) Growing rich again, *The Economist*, 9 April

Economist (1988b) Through the roof, *The Economist*, 21 May, 41

Economist (1989) When house prices fall, *The Economist*, 11 November, 20

Economist (1991) British insurers: another disaster, *The Economist*, 7 September, 111–12

Economist (1992a) An Englishman's home is his prison, *The Economist*, 18 July, 56–7

Economist (1992b) Unsafe as houses, *The Economist*, 8 August, 41–2

Economist (1993) I own, I owe, so off to work I go, *The Economist*, 8 January, 95–7

Economist (1997) Another housing boom, *The Economist*, 18 January, 27–8

Edel, M. (1981) Home ownership and working class unity, *International Journal of Urban and Regional Research*, 6, 205–22

Edel, M., Sclar, E. and Luria, D. (1984) *Shaky Places*, New York, Columbia University Press

Engels, F. (1969 edn) The housing question, in Marx, K. and Engels, F. *Selected Works*, vol. 2, London, Lawrence and Wishart

Ermisch, J. (1990) (ed.) *Housing and the Wider Economy*, Aldershot, Avebury

Ermisch, J. (1991) An ageing population, household formation and housing, *Housing Studies*, 6, 4, 230–40

Evans, A.W. (1989) South East England in the eighties: explanations for a house price explosion, in Breheny, M. and Congdon, T. (eds) *Growth and Change in a Core Region*, London, Pion

Evans, A.W. (1991a) "Rabbit hutches on postage stamps", planning development and political economy, *Urban Studies*, 28, 6, 853–70

Evans, A.W. (1991b) Investment diversion and equity release: macroeconomic consequences of increasing property values, *Urban Studies*, 26, 2, 173–82

Evans, A.W. (1996) The impact of land use planning and tax subsidies on the supply and price of housing in Britain: a comment, *Urban Studies*, 33, 3, 581–5

Farmer, M. and Barrell, R. (1984) Entrepreneurship and government policy: the case of the housing market, *Journal of Public Policy*, 2, 307–32

Feinstein, C. (1992) Britain's hidden wealth revolution, *Independent*, 18 May

Fleming, M.C. and Nellis, J.G. (1983) A regional comparison of house price inflation rates in Britain 1967 – a comment, *Urban Studies*, 20, 91–5

Fleming, M.C. and Nellis, J.G. (1985) *Housing Policy and Future of Home-ownership in the UK*, School of Management, Cranfield Institute of Technology

Fleming, M.C. and Nellis, J.G. (1990) The rise and fall of house prices: causes, consequences and prospects, *National Westminster Bank Quarterly Review*, November, 34–51

Florida, R.L. and Feldman, M.A. (1988) Housing in US Fordism, *International Journal of Urban and Regional Research*, 12, 2, 187–209

Foley, P. (1986) House price boom or bust, *Lloyds Bank Economic Bulletin*, no 93, September

Foley, P. (1991a) As safe as houses? *Lloyds Bank Economic Bulletin*, no 147

Foley, P. (1991b) Housing recovery: blessing or curse, *Lloyds Bank Economic Bulletin*, no 149, May

Foley, P. (1992) The savings puzzle, *Lloyds Bank Economic Bulletin*, September

Ford, J. (1993) Mortgage possession, *Housing Studies*, 8, 4, 227–40

Ford, J. (1995) A changing labour market: the context for mortgage lending in the 1990s, *Housing Finance*, 28, 6–13

Forrest, R. (1983) The meaning of home ownership, *Environment and Planning A: Society and Space*, 1, 205–16

Forrest, R. and Kemeny, J. (1982) Middle class housing careers – relationship between furnished renting and home ownership, *Sociological Review*, 30, 208

Forrest, R. and Kennett, P. (1996) Coping strategies, housing careers and households with negative equity, *Journal of Social Policy*, 25, 3, 369–94

Forrest, R. and Leather, P. (1995) The future of home ownership, *Housing Finance*, 27, 9–15

Forrest, R. and Murie, A. (1986) Marginization and subsidized individualism, *International Journal of Urban and Regional Research*, 10, 46–66

Forrest, R. and Murie, A. (1987a) Fiscal reorientation, centralisation and the privatization of council housing, in van Vliet, W. (ed.) *Housing Markets and Policies under Fiscal Austerity*, Westport, Greenwood

Forrest, R. and Murie, A. (1987b) The affluent home owner: labour market position and the shaping of housing histories, *Sociological Review*, 35, 2, 370–403

Forrest, R. and Murie, A. (1987c) Spatial mobility, tenure mobility and emerging social divisions in the UK housing market, *Environment and Planning*, 19, 167–30

Forrest, R. and Murie, A. (1988) *Selling the Welfare State: the privatisation of public housing*, London, Routledge

Forrest, R. and Murie, A. (1989a) Differential accumulation: wealth, inheritance and housing policy reconsidered, *Policy and Politics*, 17, 1, 25–39

Forrest, R. and Murie, A. (1989b) Housing markets, labour markets and housing histories, in Allen, J. and Hamnett, C. (eds) *Housing Markets and Labour Markets*, London, Unwin Hyman

Forrest, R. and Murie, A. (1990a) A dissatisfied state? Consumer preferences and council housing in Britain, *Urban Studies*, 27, 5, 617–35

Forrest, R. and Murie, A. (1990b) *Residualisation and council housing*, School for Advanced Urban Studies, University of Bristol, Working Paper 91

Forrest, R. and Murie, A. (1990c) *Moving the Housing Market: council estates, social change and privatization*, Aldershot, Avebury

Forrest, R. and Murie, A. (1994) Home ownership in recession, *Housing Studies*, 8, 4, 227–40

Forrest, R. and Murie, A. (1995) From privatization to commodification: tenure conversion and new zones of transition in the city, *International Journal of Urban and Regional Research*, 19, 3, 407–22

Forrest, R., Gordon, D. and Murie, A. (1996) The position of former council homes in the housing market, *Urban Studies*, 33, 1, 125–36

Forrest, R., Murie, A. and Williams, P. (1990) *Home Ownership: differentiation and fragmentation*, London, Unwin Hyman

Forrest, F., Kennett, P., Leather, P. and Gordon, D. (1994) *Home Owners in Negative Equity*, School for Advanced Urban Studies, University of Bristol

Frosztega, M. (1993) Inheritance of house property, *Economic Trends*, 481, 111–19

Gentle, C., Dorling, D. and Cornford, J. (1992) The crisis in housing: disaster or opportunity, *Centre for Urban and Regional Development Studies Working Paper*, no 96, University of Newcastle

Gentle, C., Dorling, D. and Cornford, J. (1994) Negative equity in Britain: causes and consequences, *Urban Studies*, 34, 2, 181–99

German, C. (1995) Heseltine buoyant over house prices, *Independent*, 29 December

Ginsburg, N. (1983) Home ownership and socialism in Britain: a bulwark against Bolshevism, *Critical Social Policy*, 7, 34–53

Glass, R. (1963) Introduction, *London: Aspects of Change*, London, Centre for Urban Studies

Good, F.J. (1990) Estimates of the distribution of personal wealth, *Economic Trends*, 444, October, 137–57

Gray, P.G. (1947) *The British Household, The social survey*, London, Central Office of Information

Grebler, L. and Mittelbach, F.G. (1979) *The Inflation of House Prices: its extent, causes and consequences*, Washington DC, D.C. Heath

Haddon, R. (1970) A minority in a welfare state society: location of West Indians in the London housing market, *New Atlantis*, 1, 2

Halifax Building Society (1992) *The UK Housing Market: from recession to recovery*, Viewpoint: Halifax Occasional Bulletin, Autumn

Halifax Building Society (1993) *Review of the UK Housing Market: recovery set to continue*, Halifax, 29 December

Hamnett, C. (1982) Owner occupation in the 1970s: ownership or investment? *Estates Gazette*, 19 June

Hamnett, C. (1983) Regional variations in house prices and house price inflation, *Area*, 15, 2, 97–109

Hamnett, C. (1984) Housing the two-nations: socio-tenurial polarisation in England and Wales, 1961–81, *Urban Studies*, 43, 389–405

Hamnett, C. (1987a) *Accumulation, access and inequality: the owner occupied housing market in Britain in the 1970s and 1980s*, paper given at 6th Urban Change and Conflict conference, University of Kent, 20–23 September

Hamnett, C. (1987b) Conservative government housing policy in Britain, 1979–85: economics or ideology? in van Vliet, W. (ed.) *Housing Markets and Policies under Fiscal Austerity*, Westport, Greenwood

Hamnett, C. (1989a) Consumption and class in contemporary Britain, in Hamnett, C., McDowell, L. and Sarre, P. (eds) *The Changing Social Structure*, London, Sage

Hamnett, C. (1989b) The owner occupied housing market in Britain: a north–south divide, in Lewis, J. and Townsend, A. (eds) *The North–South Divide*, London, Paul Chapman

Hamnett, C. (1989c) The social and spatial segmentation of the London owner occupied housing market: an analysis of the flat conversion sector, in Breheny, M. and Congdon, T. (eds) *Growth and Change in a Core Region*, Pion

Hamnett, C. (1991) A nation of inheritors? Housing inheritance, wealth and inequality in Britain, *Journal of Social Policy*, 20, 4, 509–36

Hamnett, C. (1992a) The geography of housing wealth and inheritance in Britain, *The Geographical Journal*, 158, 307–21

Hamnett, C. (1992b) House-price differentials, housing wealth and migration, in Champion, T. and Fielding, T. (eds) *Migration Processes and Patterns*, London, Belhaven

Hamnett, C. (1993a) The spatial impact of the British home ownership market slump, 1989–91, *Area*, 25, 217–27

Hamnett, C. (1993b) Running housing policy and the British housing system, in Maidment, R. and Thompson, G. (eds) *Managing the United Kingdom*, London, Sage

Hamnett, C. (1994) Restructuring housing finance and the housing market, in Corbridge, R., Thrift, N. and Martin, R. (eds) *Money, Power and Space*, Oxford, Blackwell

Hamnett, C. (1995a) Housing inheritance and inequality: a response to Watt, *Journal of Social Policy*, 24, 3, 413–22

Hamnett, C. (1995b) Housing inheritance and wealth in Britain, in Walker, A. (ed.) *The New Generational Contract*, London, UCL Press

Hamnett, C. (1995c) Housing equity release and inheritance, in Allen, J. and Perkins, L. (eds) *The Future of Family Care for Older People*, London, HMSO

Hamnett, C. (1996) Housing and the middle classes, in Butler, T. and Savage, M. (eds) *The Middle Classes in Britain*, London, UCL Press

Hamnett, C. (1997a) A stroke of the Chancellor's pen: the social and regional impact of the Conservatives' 1988 higher rate income tax cuts, *Environment and Planning A*, 29, 129–47

Hamnett, C. (1997b) Housing wealth, inheritance and residential care in Britain, *Housing Finance*, 34, 35–8

Hamnett, C. and Randolph, W. (1987) The residualisation of council housing in Inner London, in Clapham, D. and English, J. (eds) *Public Housing: current trends and future developments*, London, Croom Helm

Hamnett, C. and Randolph, W. (1988) *Cities, Housing and Profits*, London, Hutchinson

Hamnett, C. and Seavers, J. (1994a) *The Geography of House Prices and House Price Inflation in the South East of England*, The South East Programme, OP 12, The Open University

Hamnett, C. and Seavers, J. (1994b) *A Step on the Ladder: home ownership careers in the South East*, The South East Programme, OP 15, The Open University

Hamnett, C. and Seavers, J. (1995) *Winners and Losers: the distribution of capital gains and losses from home ownership in the South East of England*, South East Programme, OP 16, The Open University

Hamnett, C. and Seavers, J. (1996) Home ownership, housing wealth and wealth distribution in Britain, in Hills, J. (ed.) *New Inequalities: the changing distribution of income and wealth in the UK*, Cambridge, Cambridge University Press

Hamnett, C. and Williams, P. (1993) Housing wealth and inheritance, in McLennan, D. and Gibb, K. (eds) *Housing Finance and Subsidies in Britain*, Aldershot, Avebury

Hamnett, C., Harmer, M. and Williams, P. (1991) *Safe as Houses: housing inheritance in Britain*, London, Paul Chapman

Hansard vol 967, cols 79–80, HMSO

Harbury, C.D. (1962) *Inheritance and the Distribution of Personal Wealth in Britain*, London, Allen and Unwin

Harbury, C.D. and Hitchens, D.M. (1979) *Inheritance and Wealth Inequality in Britain*, London, Allen and Unwin

Harloe, M. (1984) Sector and class: a critical comment, *International Journal of Urban and Regional Research*, 8, 2, 228–37

Harmer, M. and Hamnett, C. (1990) Regional variations in housing inheritance in Britain, *Area*, 22, 1, 5–15

Harris, R. (1986) Boom and bust: the effects of house price inflation on home ownership patterns in Montreal, Toronto, and Vancouver, *The Canadian Geographer*, 30, 4, 302–15

Harris, R. and Hamnett, C. (1987) The myth of the promised land: the social diffusion of home ownership in Britain and North America, *Annals of the Association of American Geographers*, 77, 2, 173–90

Harvey, D. (1974) Class monopoly rent, finance capital and the urban revolution, *Regional Studies*, 8, 239–55

Hay, C. (1992) Housing policy in transition: from the post-war settlement towards a "Thatcherite" hegemony, *Capital and Class*, 46, Spring, 27–64

Henderson, J. and Karn, V. (1984) Race, class and the allocation of public housing in Britain, *Urban Studies*, 21, 115–28

Hendry, D. (1983) Econometric modelling: the consumption function in retrospect, *Scottish Journal of Political Economy*, 30, 193–220

Hendry, D. (1984) Economic modelling of house prices in the UK, in Hendry, D. and Wallis, K. (eds) *Econometrics and Quantitative Economics*, Oxford, Blackwell

Heseltine, M. (1979) *House of Commons Debates*, 15 May, vol 967, col 80, London, HMSO

Hills, J. (1992) *Unravelling Housing Finance: subsidies, benefits and taxation*, Oxford, Clarendon Press

Hirayama, Y. and Hayakawa, K. (1995) Home ownership and family wealth in Japan, in Forrest, R. and Murie, A. (eds) *Housing and Family Wealth: comparative international perspectives*, London, Routledge

Hirsch, F. (1978) *The Social Limits to Growth*, London, Routledge and Kegan Paul

Hogarth, R., Elias, P. and Ford, J. (1996) *Mortgages, Families and Jobs: the growth of home ownership in the 1980s*, Institute of Employment Research, University of Warwick

Holmans, A.E. (1986) *Flows of funds associated with house purchase for owner occupation in the United Kingdom, 1977–1984 and equity withdrawal from house purchase finance*, Government Economic Service Working Paper no 92, London, Department of the Environment

Holmans, A.E. (1987) *Housing Policy in Britain*, London, Croom Helm

Holmans, A.E. (1990) *House Prices: changes through time at the national and subnational level*, Government Economic Service Working Paper no 110, London, Department of the Environment

Holmans, A.E. (1991) *Estimates of housing equity withdrawal by owner occupiers in the United Kingdom 1970 to 1990*, Government Economic Service Working Paper no 116, Department of the Environment

Holmans, A.E. (1995) Where have all the first-time buyers gone? Estimating the first-time buyer population in the UK, *Housing Finance*, 25, 7–13

Holmans, A.E. (1996) A decline in young owner-occupiers in the 1990s, *Housing Finance*, 29, 13–21

Holmans, A.E. (1997) Housing and inheritance revisited, *Housing Finance*, 33, 14–23

Holmans, A. and Frostega, M. (1994) *House Property and Inheritance in the UK*, London, HMSO

Hughes, J.W. (1993) Clashing demographics: home ownership and affordability dilemmas, *Housing Policy Debate*, 2, 4, 1217–50

Huhne, C. (1992) Killing coal digs a pit for economy, *Independent*, 18 October

Hutton, W. (1991) Housing is where recovery is, *Guardian*, 23 September

Hutton, W. (1995) Revolution that casts into political stone fear of losing our homes, *Guardian*, 22 May

Independent (1987) MP's fury at £36,500 to live in cupboard, *Independent*, 16 February

Independent (1995) Tory fears of repossession, *Independent*, 6 June

Ineichen, B. (1981) The housing decisions of young people, *British Journal of Sociology*, 32, 252–68

James, N. (1993) Inheritance: all our futures, *Lloyds Bank Economic Bulletin*, 176, August

James, S., Jordan, B. and Kay, H. (1991) Poor people, council housing and the right to buy, *Journal of Social Policy*, 20, 1, 27–40

Jeager, M. (1986) Class definition and the aesthetics of gentrification, in Smith, N. and Williams, P. (eds) *Gentrification of the City*, London, Allen and Unwin

Jennings, J.H. (1971) Geographical implications of the municipal housing programme in England and Wales, 1919–1939, *Urban Studies*, 8, 121–38

Johnston, C. (1986) House price boom or bust? *Lloyds Bank Economic Review*, 93, September

Johnston, R.J. (1987) A note on housing tenure and voting, *Housing Studies*, 2, 2, 112–121

Jones, C. (1978) Household movement, filtering and trading up within the owner occupied sector, *Regional Studies*, 12, 551–61

Kaletsky, A. (1995) No house room for the myth of negative equity, *The Times*, 22 June

Karn, V., Kemeny, J. and Williams, P. (1985) *Home Ownership in the Inner City – salvation or despair*, London, Croom Helm

Kelly, R. (1992) Home ownership market hits the buffers, *The Times*, 8 April

Kemeny, J. (1981) *The Myth of Home Ownership: private versus public choices in housing tenure*, London, Routledge

Kemeny, J. and Thomas, A. (1984) Capital leakage from owner occupied housing, *Policy and Politics*, 12, 1, 13–30

Kemp, P. (1982) Housing landlordism in late nineteenth century Britain, *Environment and Planning A*, 14, 1437–47

Kemp, P. (1990) Shifting the balance between state and market: the reprivatisation of rental housing provision in Britain, *Environment and Planning A*, 22, 793–810

Kendig, H. (1984a) Housing careers, life cycle and residential mobility: implications for the housing market, *Urban Studies*, 21, 271–83

Kendig, H. (1984b) Housing tenure and generational equity, *Ageing and Society*, 4, 3, 249–72

Kessler, D. and Wolff, E.N. (1991) A comparative analysis of household wealth patterns in France and the United States, *Review of Income and Wealth*, 37, 3, 249–66

Kilroy, B. (1979) Housing finance: why so privileged, *Lloyds Bank Review*, 133, June, 37–52

Laing and Buisson (1991) Laing and Buisson Annual Review

Lauria, D. (1976) Wealth, capital and power: the social meaning of home ownership, *Journal of Interdisciplinary History*, 7, 2, 261–82

Leather, P. (1990) The potential and implications of home equity release in old age, *Housing Studies*, 5, 1, 3–13

Leather, P. and Williams, P. (1996) Renewal of the private housing stock: a review of longer term prospects, *Housing Finance*, 31, 33–40

Lee, C. and Robinson, B. (1989a) Can household surveys help explain the fall in the savings ratio? *Fiscal Studies*, 10, 3, 58–71

Lee, C. and Robinson, B. (1989b) *Savings and Housing: Lessons from Micro Data*, Micro to Macro Monograph no 1, London, Institute for Fiscal Studies

Lee, C. and Robinson, B. (1990) The fall in the savings ratio: the role of housing, *Fiscal Studies*, 11, 1, 36–52

Leigh Pemberton, R. (1979) Bank, building societies and personal savings, *National Westminster Quarterly Bank Review*, May, 2–10

Leigh Pemberton, R. (1986) Structural change in housing finance, *Bank of England Quarterly Bulletin*, December, 26, 4, 528–31

Leigh Pemberton, R. (1991) Evidence to the Treasury and Civil Service Committee, House of Commons, April

Leigh Pemberton, R. (1993) The role of property in our economic life, *Bank of England Quarterly Bulletin*, February, 33, 1, 106–9

Levin, E.J. and Wright, R. (1997) Speculation in the housing market, *Urban Studies*, 34, 9, 1419–38

Lomax, D. (1982) The banks and the housing market, *National Westminster Quarterly Bank Review*, February, 2–12

Lomax, J. (1991) Housing finance – an international perspective, *Bank of England Quarterly Bulletin*, February, 31, 1, 55–66

Longley, P. and Williams, H. (1993) Models of trading behaviour and accumulation in stratified housing markets, *Environment and Planning A*, 67–80

Lowe, S. (1987) New patterns of wealth: the growth of owner occupation, in Walker, R. and Parker, G. (eds) *Money Matters*, London, Sage

Lowe, S. (1989) From first time buyers to last-time sellers: an appraisal of the social and economic consequencs of equity withdrawal from the housing market between 1982 and 1988, Department of Social Policy, University of York

Lowe, S. (1992) The social and economic consequences of the growth in home ownership, in Birchall, J. (ed.) *Housing Policy in Britain*, London, Routledge

Lui, Tai-lok (1995) Coping strategies in a booming market: family wealth and housing in Hong Kong, in Forrest, R. and Murie, A. (eds) *Housing and Family Wealth: comparative international perspectives*, London, Routledge

MacDonald, R. and Taylor, M.P. (1993) Regional house prices in Britain: long-run relationships and short-run dynamics, *Scottish Journal of Political Economy*, 40, 1, 43–55

McLaverty, P. and Yip, N.M. (1993) The preference for owner occupation, *Environment and Planning A*, 25, 1559–72

McLaverty, P. and Yip, N.M. (1994) Income multiples and mortgage potential, *Urban Studies*, 31, 8, 1367–76

MacLennan, D. and Gibb, K. (1990) Housing finance and subsidies in Britain after a decade of Thatcherism, *Urban Studies*, 27, 6, 905–18

MacLennan, D. and Gibb, K. (1993) Housing and the UK economy: the boom, the bust and beyond, in D. MacLennan and K. Gibb: *Housing Finance and Subsidiaries in Britain*, Avebury, Aldershot

McRae, H. (1992a) Signs of recovery for the greater good, *Independent*, 22 July

McRae, H. (1992b) Paying for the future with hard saved cash, *Independent*, 10 June

McRae, H. (1993) Home, no longer where the money is, *Independent*, 3 December

Malpass, P. (1990) *Reshaping Housing Policy: subsidies, rents and residualisation*, London, Routledge

Malpass, P. and Murie, A. (1990) *Housing Policy and Practice*, London, Macmillan

Marfleet, T. and Pannell, B. (1996) House prices and affordability, *Housing Finance*, 31, 6–16

Meen, G.P. (1989) The ending of mortgage rationing and its effects on the housing market: a simulation study, *Urban Studies*, 26, 240–52

Meen, G.P. (1995) Is housing good for the economy, *Housing Studies*, 10, 3, 405–24

Meen, G.P. (1996) Ten propositions in UK housing macroeconomics: an overview of the eighties and early nineties, *Urban Studies*, 33, 3, 425–44

Merrett, S. and Gray, F. (1982) *Owner Occupation in Britain*, London, Routledge

Miles, D. (1992a) Housing and the wider economy in the short and long run, *National Institute Economic Review*, February, 64–77

Miles, D. (1992b) Housing markets, consumption and financial liberalisation in the major economies, *European Economic Review*, 36, 1093–1136

Milne, A. and Clark, A. (1990) House prices, housing investment and demography, Forecasts from the LBS/Abbey National Model of the UK Housing Market, LBS Briefing Paper, London

Moon, P. (1992) Housing gives less shelter, *Lloyds Bank Economic Bulletin*, no 157, January

Moon, P. (1993) Housing on a better footing, *Lloyds Bank Economic Bulletin*, no 172, April

Moon, P. (1994) New stability for housing, *Lloyds Bank Economic Bulletin*, no 189, September

Moon, P. (1995) Housing detached from recovery, *Lloyds Bank Economic Bulletin*, no 3, June

Morgan Grenfell (1987) *Housing Inheritance and Wealth*, Morgan Grenfell Economic Review, 45, November

Morgan Grenfell (1989) *Housing Slump – the next phase*, UK Economic Issues, February

Morgan Grenfell (1992) *The Housing Slump: when will it end?*, London, Morgan Grenfell Economics

Morgan Grenfell (1993) *Housing Inheritance: how much, how soon?* UK Economic Issues, July

Muellbauer, J. (1990a) The housing market and the UK economy, problems and opportunities, in Ermisch, J. (ed.) *Housing and the National Economy*, Aldershot, Avebury

Muellbauer, J. (1990b) The great British housing disaster, *Roof*, May/June

Muellbauer, J. (1990c) *The Great British Housing Disaster and Economic Policy*, Economic Study no 5, London, Institute for Public Policy Research

Muellbauer, J. (1991) Anglo-German differences in housing market dynamics: the role of institutions and macro economic policy, Nuffield College Oxford

Muellbauer, J. (1997) X-rating the housing boom, *Observer*, 4 May

Muellbauer, J. and Murphy, A. (1991) Regional economic disparities: the role of housing, in Bowen, A. and Mayhew, K. (eds) *Reducing Regional Inequalities*, London, NEDO

Munro, M. (1989) Housing wealth and inheritance, *Journal of Social Policy*, 17, 4, 417–36

Murie, A. (1986) Social differentiation in urban areas: housing or occupational class at work? *Tijdschrift voor Economische en Socaile Geografie*, 77, 5, 345–57

Murie, A. (1991) Divisions of homeownership: housing tenure and social change, *Environment and Planning A*, 23, 349–70

Murie, A. and Forrest, R. (1980) Wealth, inheritance and housing policy, *Policy and Politics*, 8, 1, 1–19

Murphy, M.J. (1984) The influence of fertility, early housing-career and socio-economic factors on tenure determination in contemporary Britain, *Environment and Planning A*, 16, 1303–18

Nationwide Building Society (1976) *Occasional Bulletin*, no. 134

Nationwide Building Society (1985a) *House Prices over the past Thirty Years*

Nationwide Building Society (1985b) *Lending to Women*, February

Nationwide Building Society (1986) *Housing as an Investment*, April

Nationwide Building Society (1987) *House Prices: North–South Divide*, August

Nationwide Building Society (1995) 1.5 million households in negative equity, *Housing Finance Review*, 2

Nationwide Building Society (1997a) Recovery broadens, but no boom, *Housing Finance Review*, 10, July

Nationwide Building Society (1997b) *Housing Finance Review*, 11, October

Nationwide Building Society (1998) Has the housing market peaked? *Housing Finance Review*, 12, January

Nicholson-Lord, D. (1993) Londoners caught in negative equity trap, *Independent*, 25 November

Nordvik, V. (1995) Prices and price expectations in the market for owner occupied housing, *Housing Studies*, 10, 3, 365–80

Pahl, R. (1975) *Whose City?* London, Penguin

Pannell, B. (1990) Trends in the personal sector balance sheet, *Housing Finance*, 8, 17–19

Pannell, B. (1992) The outlook for personal sector borrowing, *Housing Finance*, 13, 6–9

Pannell, B. (1996) Moving owner occupiers, *Housing Finance*, 32, 16–20

Pannell, B. (1997) Tenure choice and mortgage decisions: 1996 market research findings, *Housing Finance*, 33, 9–13

Pannell, B. and Abisogun, L. (1992) House prices and earnings, *Housing Finance*, 15, 9–18

Parsons, D. (1987) Recruitment difficulties and the housing market, *The Planner*, January, 30–34

Pattie, C., Dorling, D. and Johnston, R. (1995) A debt-owing democracy: the political impact of housing market recession at the British general election of 1992, *Urban Studies*, 32, 8, 1293–1316

Pawley, M. (1975) Playing reverse monopoly with houses, *Guardian*, 26 July

Pawley, M. (1978) *Home Ownership*, London, The Architectural Press

Pawley, M. (1985) Playing reverse monopoly with housing, *Guardian*, 26 June

Payne, J. and Payne, G. (1977) Housing pathways and stratification: a study of life chances in the housing market, *Journal of Social Policy*, 6, 129–56

Peach, C. and Byron, M. (1993) Caribbean tenants in council housing, "race", class and gender, *New Community*, 19, 407–23

Peck, J. and Tickell, A. (1992) Local modes of social regulation? Regulation theory, Thatcherism and uneven development, *Geoforum*, 23, 347–63

Penycate, J. (1986) The property boom widens the north–south divide, *The Listener*, 13 November, 4–5

Peseran, M. and Evans, R. (1984) Inflation, capital gains and UK personal savings, 1953–81, *Economic Journal*, 94, 237–57

Pickles, A.R. and Davies, R.B. (1991) The empirical analysis of housing careers: a review and a general statistical modelling framework, *Environment and Planning A*, 23, 465–84

Pickvance, C. and Pickvance, K. (1994) Towards a strategic approach to housing behaviour, *Sociology*, 28, 3, 657–77

Pratt, G. (1982) Class analysis and urban domestic property: a critical re-examination, *International Journal of Urban and Regional Research*, 6, 481–502

Purves, L. (1995) Home, soured home, *The Times*, 26 September

Ranney, S. (1981) The future of house prices, mortgage market conditions and the returns to home ownership, *American Economic Review*, 71, 323–33

Rathbone, J. (1992) How Town and Country bit the dust, *Observer*, 17 May

Raynor, W. (1997) The boom that never was, *Independent on Sunday*, 30 November

Reidy, M. (1994) Inequalities in housing tenure attainment in Britain, D. Phil. thesis, Nuffield College, University of Oxford

Rentoul, J. (1996) Blair lays claim to home owners' votes, *Independent*, 6 March

Revell, J.R.S. (1967) *The Wealth of the Nation*, Cambridge, Cambridge University Press

Riley, B. (1992) Mortgages that threaten ruin, *Financial Times*, 17 October

Robinson, R., O'Sullivan, T. and Le Grand, J. (1985) Inequality and housing, *Urban Studies*, 22, 249–56

Rosenthal, L. (1989) Income and price elasticities of demand for owner-occupied housing in the UK, *Applied Economics*, 21, 761–75

Rowe, W.M. and Stegman, M.A. (1994a) The effects of homeownership on the self-esteem, perceived control and life satisfaction of low-income people, *American Planning Association Journal*, April, 60, 2, 173–84

Rowe, W.M. and Stegman, M.A. (1994b) The impact of home ownership on the social and political involvement of low-income people, *Urban Affairs Quarterly*, 30, 28–50

Royal Commission on the Distribution of Income and Wealth (1977) *Third Report on the Standing Reference, no 5*, Cmd 6999, London, HMSO

Rubenstein, W.D. (1986) *Wealth and Inequality in Britain*, London, Faber and Faber

Ruonavaara, H. (1988) *The Growth of Urban Home-ownership in Finland, 1950–1980*, Department of Sociology Studies no 10, University of Turku

Ruonavaara, H. (1993) Home-owners in distress: financial deregulation and the 1990s: home owner crisis in Finland, Paper given at European Network for Housing Research Conference, Budapest, 7–10 September

Saunders, P. (1978) Domestic property and social class, *International Journal of Urban and Regional Research*, 2, 233–51

Saunders, P. (1984) Beyond housing classes: the sociological significance of private property rights in means of consumption, *International Journal of Urban and Regional Research*, 8, 202–27

Saunders, P. (1986) Comment on Dunleavy and Preteceille, *Environment and Planning A*, 4, 155–63

Saunders, P. (1990) *A Nation of Home Owners*, London, Unwin Hyman

Saunders, P. and Harris, C. (1988) *Home Ownership and Capital Gains*, WP 64, Urban and Regional Studies, University of Sussex

Saunders, P. and Williams, P. (1988) The constitution of the home, *Housing Studies*, 3, 1, 81–93

Savage, M., Watt, P. and Arber, S. (1990) The consumption sector debate and housing mobility, *Sociology*, 24, 1, 97–117

Savage, M., Barlow, J., Dickens, P. and Fielding, T. (1992) *Property, Bureaucracy and Culture: middle class formation in contemporary Britain*, London, Routledge

Sheffman, D.T. (1978) Some evidence on the recent boom in land and housing prices, in Bourne, L. and Hitchcock, J.K. (eds) *Urban Housing Markets: recent directions in research and policy*, University of Toronto

Simpson, M.A. and Lloyd, T.H. (1977) *Middle Class Housing in Britain*, Newton Abbott, David and Charles

Smith, D. (1992) *From Boom to Bust: trial and error in British economic policy*, London, Penguin

Spackman, A. (1997) Back to the bijou bubble of '87, *Financial Times*, 11 October

Spencer, P. (1987) *U.K. House Prices – not an inflation signal*, Credit Suisse First Boston Economics, September

Spencer, P. and Scott, A. (1990) *Credit controls, Inflation and Spending: a new perspective on the behaviour of consumer expenditure in the UK*, London, Shearson Lehman Hutton, 25 January

Stephens, M. (1993a) Responses to recession: the housing slump in Britain, Paper given at European Network for Housing Research Conference, Budapest, 7–10 September

Stephens, M. (1993b) Housing finance deregulation: Britain's experience, Netherlands *Journal of Housing and the Built Environment*, 8, 2, 159–75

Stephens, M. (1993c) Finance for owner occupation in the UK: the sick man of Europe, *Policy and Politics*, 21, 4, 307–17

Stephens, M. (1995) Monetary policy and house price volatility in western Europe, *Housing Studies*, 10, 551–64

Stephens, M. (1996) Institutional responses to the UK housing market recession, *Urban Studies*, 33, 2, 337–52

Stern, D. (1992) Explaining UK house price inflation, 1971–89, *Applied Economics*, 24, 1327–33

Sternlieb, G. and Hughes, J. (1972) The post-shelter society, *The Public Interest*, 39–47

Sullivan, O. and Murphy, M. (1987) Young outright owner occupiers in Britain, *Housing Studies*, 2, 3, 177–91

Swennarton, M. and Taylor, S. (1985) The scale and nature of the growth of owner occupation in Britain between the wars, *Economic History Review*, 38, 3, 373–92

Symonds, B. (1989) Relationship breakdown and the housing moves of owner occupiers, in McCrthy, P. and Simpson, B. (eds), *Issues in Post Divorce Housing*, Aldershot, Avebury

Thatcher, M. (1979) *Debate on the Queen's Speech*, Hansard London, HMSO, May

Thomas, R. (1996) The sun shines on houses, *Financial Times*, 20 October

Thorns, D.C. (1981a) Owner occupation: its significance for wealth transfer and class formation, *Sociological Review*, 29, 4, 705–27

Thorns, D.C. (1981b) The implications of differential rates of capital gain from owner occupation for the formation and development of housing classes, *International Journal of Urban and Regional Research*, 5, 205–17

Thorns, D.C. (1989) The impact of homeownership and capital gains upon class and consumption sectors, *Environment and Planning D, Society and Space*, 7, 293–312

Thorns, D.C. (1992) *Fragmenting Societies*, Auckland NZ, Routledge

Thorns, D.C. (1994) The role of housing inheritance in selected owner occupied societies (Britain, New Zealand and Canada), *Housing Studies*, 9, 4, 473–92

Thrift, N. and Leyshon, A. (1991) In the wake of money, in Budd, L. and Whimpster, S. (eds) *Global Finance and Urban Living*, London, Routledge

Thurow, L. (1969) The optimal lifetime distribution of consumption expenditure, *American Economic Review*, 59, 324–30

Timmins, N. (1992) Home owners "let down" by Tories, *Independent*, 31 January

Troy, P.N. (1991) *The Benefits of Owner Occupation*, WP 29 Urban Research Programme, Research School of Social Sciences, Canberra, Australian National University

Tucillo, J. (1980) Housing and investment in an inflationary world: theory and evidence, Washington DC, The Brookings Institute

Turnbull, P. (1990) *The Housing Market: outlook and analysis*, London, Smith New Court, 5 October

Turner, B. (1993) Swedish home owners: crises and defaults, Paper given at European Network for Housing Research Conference, Budapest, 7–10 September

UBS Phillips and Drew (1991) *Banks and Building Societies – bloodbath in the High Street*, UBS Phillips and Drew Global Research Group, February

UBS Phillips and Drew (1992) *Housing Market: more pain before gain*, UBS Phillips and Drew Global Research Group, November

Valley, P. (1995) The house that Thatcher dismantled, *Independent,* 17 February

van der Weyer, M. (1994) The age of the last-time seller, *Spectator*, 17 September

van der Weyer, M. (1995) A house is not a goldmine, *Independent*, 22 July

Waldegrave, W. (1987) *Some reflections on housing policy*, Conservative News Service, London

Walker, R. ((1981) A theory of suburbanization: capitalism and the construction of urban space in the United States, in Dear, M. and Scott, A. (eds) *Urbanization and Urban Planning in Capitalist Society*, London, Methuen

Watt, P. (1993) Housing inheritance and social inequality: a rejoinder to Chris Hamnett, *Journal of Social Policy*, 22, 4, 527–34

Webster, P. and Lumsden, G. (1988) Minister warns banks over return of 100% mortgages, *The Times*, 2 February

Westaway, P. (1993) *Mortgage Equity Withdrawal: causes and consequences*, National Institute of Economic and Social Research Discussion Paper 59

Westaway, P. (1994) *Mortgage Equity Withdrawal: causes and consequences*, Housing Research Findings no 119, York, Joseph Rowntree Foundation

References

Whitehead, C. (1979) Why owner occupation? *CES Review*, May, 33–41, London, Centre for Environmental Studies

Wiener, M. (1981) *English Culture and the Decline of the Industrial Spirit, 1850–1980*, Cambridge, Cambridge University Press

Wood, D. (1995) Twenty years of falling prices, *Independent*, 23 May

Wood, D. (1996) But here comes the night, *Financial Times*, 20 October

Wray, M. (1968) Building society mortgages and the housing market, *Westminster Bank Quarterly Review*, February, 31–45

Wriglesworth, J. (1992) Housing market: the debt trap, *UBS Phillips and Drew Economic Briefing*, 262, 9 June

Index

Note: References are to **housing** and **home ownership** and **Britain**, unless otherwise specified e.g. investment is *investment, housing as*

Abbey National 38, 41, 42
Abelson, P. 21
Abisogun, L. 36, 38
accumulation *see* equity; wealth
age 124
    equity distribution 111–12, 113, 118, 120
    investment strategies 156, 159–60
    see also first-time; last-time
Agnew, J. 147–8
Anderson, M. 152
arrears and repossessions 17, 22, 36–8, 121, 155, 172
Atkinson, A.B. 103–4
Australia 21, 51, 57, 61, 75, 124, 146, 185

baby boom 24, 32, 46, 186, 188
Badcock, B. 21, 76
Ball, M. 66, 68
Bank of England 16
    equity 110, 134–5
        negative 115–16
    investment 73, 76
    wealth and economy 169–70, 175–6, 178, 181
Barber, A. 24, 47, 177
Barrell, R. 147, 149
Beaverstock, J. 41–2
Bellman, H. 67
Bentham, G. 59
BHPS *see* British Household Panel Survey
Black Horse Agencies 41, 42
Blair, C. 11

Blair, T. 9–11, 12
Boleat, M. 167
booms 1–5, 12, 13, 101–2, 183, 191–2
    of 1970s 17, 24–5, 47, 177
    of 1980s 17, 24, 27, 29–32, 139–40, 193–4
    wealth and economy 167–71, 177–81
    expected 151
    triggers for 46–8
Bootle, R. 9, 48, 184
borrowing 173, 175–6, 178–9
    see also lending institutions
Bourdieu, P. 55, 61
Bover, O. 180
Bowen, A. 167
Bramley, G. 190
Breedon, F.J. 22, 24, 38, 46, 47, 48, 186
Brindley, T. 147
British Household Panel Survey
    class 62, 63
    equity 107–8, 109–14
    gains and losses 78–81
British Property Federation 76
British Social Attitudes Survey 45, 159, 196
building new houses 17, 27, 52, 56, 57, 167, 192
building societies *see* lending institutions
Building Societies Act (1986) 41
Building Societies Association 22, 24, 34–5, 49, 174
Burnett, J. 56
Burrows 196
Byron, M. 64

217

Canada 21, 51, 57, 185
capital *see* equity
careers *see* strategies *under* investment
Carruth, A. 167, 173–4, 181
Case, K. 21, 193
Census (91) 108
chains and equity extraction 136
Checkoway, B. 167
Chrystal, K.A. 191
Cicutti, N. 42
Clapham, D.
  future 191, 194
  social basis 54
  structure of market 39, 47
  wealth and economy 167, 178, 179
Clark, A. 24, 47, 186, 192
class 55–67
  accumulation *see* wealth
  diffusion down 55–7, 58
  equity distribution 111–12, 114, 117,
    118–19, 120
  formation 65–7
  and house types 61–4
  and income 55–7, 84–5, 160–1
    gains and losses 86–90, 95–9
  inheritance 123–5, 130–1
  investment strategies 147–8, 149, 151,
    156, 161
Coles, A. 167
confidence, erosion of 191
Congdon, T. 69
Conservatives 39, 73, 105, 134, 184
  class 51–3, 54, 61, 67–8, 69, 73, 178,
    181
consumption
  consumer spending 168–70, 174–8
    wealth and economy 167–77 *passim*
  good, housing as 145–50
  location 66
Cooper, C. 55
Cornerstone 42
Costello, J. 167, 174, 176, 177–8
council housing and tenants 130
  bought ('Right to Buy') 6, 51, 53–4,
    57, 73, 85, 185
  higher rents 54
  numbers 58–9
  voting habits 68

Council for Mortgage Lenders 159
  class 51, 55
  structure of home ownership 20, 23–4,
    26–31, 37
  wealth and economy 169, 171, 172,
    174–5
Counsell, G. 7
'country houses' 61–2
Couper, M. 147
Crow, G. 152
current value 90–2, 108
Curtice, J. 45
Cutler, J.
  future 186, 190–1, 196
  structure of market 18, 22, 24, 26, 46,
    47, 48
  wealth and economy 167, 175, 176

date of purchase 10, 151
  equity distribution 117–18, 119, 120
  gains 92–3
  investment 73–4, 76, 85, 88–9, 92–3
Daunton, M. 12, 56
Davidson, J. 168
Davis, E.P. 134, 135, 137
demographic change 46, 185–90
  baby boom 24, 32, 46, 186, 188
Denmark 21
Department of Environment 34–5, 49,
    189–90
Deverson 147
Dicks, M.J. 179, 186
Dieleman, F. 21
Diggens, J.P. 61
Doling, J. 2, 73, 155
Donnison, D. 56
Dorling, D. 116–18
Downs, A. 74, 134, 146, 147, 189
Duncan, S. 75
Dupuis, A. 21, 76, 77–8, 85

Earley, F. 45, 64
East Anglia 115, 130
  1970s and 1980s 27, 32
  equity distribution 110–11
    negative 115, 117, 119
  investment 74, 80

recovery 43
slump of 1990s 33–6
Eatwell, J. 7, 179
economy *see under* wealth
Edel, M. 68
Eden, A. 54
Engels, F. 68
equity 2, 13, 15, 133–43
  accumulation 65–7, 74
    see also wealth, housing
  distribution 15, 101–22
    negative 115–21
    in South East 114–15
    variations in mean 110–13
    see also extraction
Ermisch, J. 24
Ermison, J. 167, 168, 177, 188
estate agents, lenders as 41–3
Europe 21, 51, 74–5, 153, 184
Evans, A.W. 17
extraction of equity 2, 13, 15
  measuring 133–43
  wealth and economy 169–70, 172–5,
    179–80

Farmer, M. 147, 149
Feinsetin, C. 105
Feldman, M.A. 167–8
finance *see* lending institutions
financial deregulation 139, 191–2
  structure of home ownership 41, 46, 47
  wealth and economy 174, 175, 177,
    178–80
Financial Services Act (1986) 41
Finland 21
first-time buyers 6, 12, 55
  equity distribution 115, 119–20
  investment strategies 153, 154, 155
  structure of home ownership 28, 32–3,
    45, 46
  wealth and economy 185–93 *passim*
Fleming, M.C. 21, 49
flexible labour market 195
Florida, R.L. 167–8
Foley, P. 48, 168–9, 171, 175, 176
Ford, J. 2, 73, 155, 195–6
Forrest, R. 6, 12, 45, 185
  equity distribution 102, 118–19

investment 75, 85
  strategies 146–7, 149, 152, 153,
    154–5
  social basis 54–60 *passim*, 63, 68
France 21
Frostega, M. 127, 133
future 13, 16, 183–97
  instability 191–7
  see also stability

gains and losses 76–99
  see also negative equity
'gazumping' 8
gender 64–5
General Accident 41, 42
General Household Survey 58–9, 62, 108
Gentle, C. 110, 116
gentrification 61, 63
German, C. 72
Germany 21, 51, 153, 184
Gibb, K. 54
Glass, R. 56
'golden postcodes' 9–10
Good, F.J. 105
government policy 39, 105, 134, 184
  social basis 51–3, 54, 61, 67–8, 69, 72,
    73, 178, 181
Gray, F. 12
Gray, P.G. 56

Halifax Building Society 21, 34, 35, 38,
  41, 43, 49
Hambro Countryside 41, 42
Hamnett, C. 24, 47, 185
  equity distribution 102, 113
  inheritance 125–6, 130–3
  investment 75, 78, 85
    strategies 152, 156
  social basis 51–2, 54, 58–9, 63, 69
  wealth and economy 175, 180, 181
Harbury, C.D. 124
Harloe, M. 66, 75
Harmer, M. 125, 130
Harris, R. 21, 51, 85
Harvey, D. 167
Hayakama, K. 21
Healey, D. 47
Henderson, J. 64

Hendry, D. 21, 48, 168
Henley, A. 167, 173–4, 181
Heseltine, M. 72, 73
Hills, J. 54
Hirayama, Y. 21
Hirsch, F. 61
Hitchens, D.M. 124
Holmans, A.E. 12, 18, 56, 76, 102
    future 186–8, 189, 192
    inheritance 127, 133, 137–43
home ownership 1–16
    booms 1–5, 13
    slumps 5–8, 13
    recovery 81–2
Hong Kong 21
housing equity withdrawal (HEW) 137–43
    see also extraction
Hughes, J. 146, 147
Huhne, C. 170, 175
Hutton, W. 69, 170–1

income/earnings 184, 194
    household 90–2
    investment strategies 156, 161
    prices 19–20, 23, 25, 28
    wealth and economy 172, 174, 175,
        177, 180
    see also under class
Ineichen, B. 152
inequality reduction 103–6
inflation 74, 190–1
    see also booms
inheritance 13, 15, 107, 123–33, 140, 143
    inheritors 129–31
    measuring 126–9
    shortfall explained 131–3
Inland Revenue Statistics
    equity distribution 102, 105–7
    inheritance 126–9, 131, 133
    structure of home ownership 31, 33–4
    see also tax
instability, future 191–7
insurance companies 41, 42
interest rates 5, 47, 174, 179, 180
international comparisons 18, 21, 124,
        168
    class 51, 57, 61
    future 185, 189, 193

investment 74–6, 85
    strategies 146, 147–8
investment 46, 73–100, 173, 183
    strategies 15, 145–65
    see also ladder
    see also gains and losses

Japan 18, 21
Jeager, M. 61
Jennings, J.H. 56
Jones, C. 134, 152, 155
Joyce, M.A. 22, 24, 38, 46, 47, 48, 186
Julien, P. 6

Kaletsky, A. 120
Kam, V. 63, 64
Kelly, R. 6–7, 8, 183
Kemeny, J. 51, 137, 152
Kemp, P. 54, 56
Kendig, H. 154
Kennett 152
Kilroy, B. 134

Labour Party 52, 53, 68, 72
ladder 45, 149, 150, 153–64
    class and income 160–1
    investment 74, 87–8
    measuring 161–4
    in South East 156–60
Laing and Buisson 133
last-time sellers 133–42 *passim*, 192
Lawson, N. 5, 125, 179, 180
Leather, P. 185
Lee, C. 24, 136–7, 168, 173, 177, 186
Leeds Permanent Building Society 38
Legal and General 41
legislation 41, 126, 132, 134, 178
Leigh, P. 76
Leigh Pemberton, R. 134, 169–70
lending institutions 6, 11, 64, 73, 139,
        194–5
    as estate agents 41–3
    inheritance 125–6, 130, 131
    ladder 154
    negative equity 116, 119–20
    recovery 43, 45
    slump of 1990s 34, 35–6, 38–41, 42,
        49

structure of home ownership 21, 22,
        24, 31–2
    wealth and economy 169–70, 171,
        174–6, 178–9, 181
    see also Halifax; Nationwide
Leyshon, A. 61
Lindsay 147
Lloyd, T.H. 56, 63
Lloyds Bank 126, 131, 171
London *see* South East
losses *see* gains and losses; negative
        equity
Lowe, S. 46–7, 137
Lui, T. 21

MacLennan, D. 54
Macmillan, H. 52
McRae, H. 170, 171
Mahon, K. 73
Major, J. 15, 125, 171
Malpass, P. 12, 42, 54
Marfleet, T. 17, 19, 43
Marx, K. 65
measuring
    equity extraction 133–43
    inheritance 126–9
    ladder 161–4
    prices 48–9
Meen, G.P. 167, 196
Merrett, S. 12
Midlands 63
    1970s and 1980s 25, 27, 29, 32
    equity distribution 110–11, 117, 120
    investment strategies 147–8
    slump of 1990s 33–6
Miles, D. 48, 134, 137, 167, 179–80, 191
Milne, A. 24, 47, 186, 192
Morgan Grenfell 125–6, 130, 131
MORI survey 78, 79, 114–15, 150
Mortgage Corporation 43
Mortgage Express 39
mortgages 130, 172, 173
    assistance to pay 6, 196
    availability and numbers 12, 24, 27–8,
        30–1, 51, 52, 59, 60, 134
    equity distribution 110–14, 115
    income and 30, 32–3
    low rates 46

relief *see under* tax
    wealth 110, 113–14
    women and 64
    see also arrears and repossessions;
        lending institutions; negative
        equity; overmortgaging
Muellbauer, J. 21, 47–8, 51, 153, 167,
        191
    wealth and economy 168, 178, 179, 180
Mulholland, F. 64
Munro, M. 102
Murie, A. 6, 12, 45, 52, 102, 185
    investment 75, 76–7, 85
        strategies 146–7, 149, 152
    social basis 54–60 *passim*
Murphy, M.J. 168, 180

National Home Loans Corporation 39
Nationwide Building Society 11, 64, 73
    ladder 154
    negative equity 116, 119–20
    recovery 43, 45
    slump of 1990s 35–6, 38, 41, 42, 49
NCW (net cash withdrawal) 134–5
negative equity/losses 5–6, 17, 22, 75,
        183, 184
    class 67, 68–9
    equity distribution 110, 114–21
    wealth and economy 171, 176
Nellis, J.G. 21, 49
net cash withdrawal 134–5
Netherlands 21, 74–5
New Zealand 21, 51, 57, 75, 76, 85, 185
Nicholson-Lord, D. 118
North 130, 148
    1970s and 1980s 24–5, 27, 29, 32
    equity distribution 108, 110–11,
        116–17, 120
    investment 74–5, 80–1, 85
    recovery 43
    slump of 1990s 33–6
Norway 21
number of houses owned 159, 160–1
    see also ladder

one-person households 64, 189
outright owners 51, 59, 60, 130
    wealth 110–14

overmortgaging 134–5, 137–8, 141–3, 169

Pahl, R. 67, 73
Pannell, B. 55
  structure of market 17, 19, 36, 38, 43
  wealth and economy 167, 169, 171, 172, 174
Pattie, C. 68–9
Pawley, M. 12, 134
Payne, J. and G. 152
Peach, C. 64
Peck, J. 180
Penycate, J. 180
personal wealth 102–3, 168–9, 172
  see also wealth, housing
Pickvance, C. and L. 152
political alignment 67–72
Porter, S. 68
Pratt, G. 66
prices 17–19
  of houses 12, 17–22, 48–9
    and income/earnings 19–20, 23, 25, 28
  see also booms; slumps
private rental *see under* rental
Prudential Property Services 42
psychological importance of home 55, 61–2
Purves, L. 72

race 64–5
Randolph, W. 24, 47, 52, 59, 185
recession 170–1
  see also slumps
recovery 43–6, 81–2, 171
regional differences 63, 74, 80, 130
  equity distribution 108, 110–11, 117, 120, 181
    negative 115, 116–17, 118, 120
  investment 74–5, 80–1, 85
    strategies 15, 147–8, 149
  recovery 43
  slump of 1990s 33–6
  South East and London 24–5, 27, 29, 32
  structure of home ownership 27, 33–6, 43
  see also South East and London

Reidy, M. 152
remortgaging 141–3
rental sector 12, 57, 60
  private (and decline of) 24, 51, 52, 54–5, 110
  public *see* council housing
Rentoul, J. 72
repossessions *see* arrears
residential care 131–3
Revell, J.R.S. 104
Riley, B. 34
Robinson, B. 24, 136–7, 168, 173, 177, 186
Royal Commission on Distribution of Income and Wealth 102–3, 104
Royal Life 41, 42
Ruonavaara, H. 21

Salomon Brothers 43
Sarre, P. 78
Saunders, P. 13–15, 19, 124, 194
  investment 74–5, 76, 79, 83–6, 89
    strategies 148–9, 150, 151, 152
  social basis 52, 55, 65–7
Savage, M. 149
Saville, I.D. 134, 135, 137
savings rate 169–70, 171, 172
Sax, E. 68
Scotland 63, 80, 130
  equity distribution 110–11, 181
    negative 117, 118, 119–20
  structure of home ownership 27, 35, 43
Scott, A. 48, 178
Seavers, J. 78, 113, 156
Shiller, R. 21, 193
Simpson, M.A. 56, 63
slumps 193
  of 1970s 25–7
  of 1990s 5–8, 13, 117, 128, 140, 183, 196–7
  class 67–72
  investment 75, 81
  mortgage arrears and repossessions 36–8
  negative equity 115–21
  structure of home ownership 17, 19, 21, 22, 32–42, 48

wealth and economy 167, 172–3, 177, 180, 181
  see also under lending institutions
social basis 13–14, 51–72
  government support 52–3
  political alignment 67–72
  race and gender 64–5
  see also class
social security 6, 54, 59–60, 195
socioeconomic groups *see* class
South East and London 27, 49, 171
  boom of 1970s 1, 24–5
  boom of 1980s 2–5, 29, 32
  class 56, 67
  equity distribution 110–11, 114–15, 181
    negative 115–18, 120
  future 192
  gains and losses 81–99
  inheritance 130
  investment 78–81
    strategies 147–8
  ladders in 156–60
  motives for purchase in 150–2
  recovery 8–11, 43–5
  slump of 1990s 5, 17, 25, 33–6, 196
  and wealth, housing and economy 180–1
South West 27, 43, 74, 115, 130
  equity distribution 110–11, 115, 117, 118
  investment strategies 15, 147–8, 149
  slump of 1990s 33–6
Spackman, A. 9
speculative behaviour 147–9
Spencer, P. 48, 75, 178
stability, forces for 185–92
Stafford 155
Stamp Duty 39
Stephens, P. 21, 38–9, 41
Stern, D. 46
Sternlieb, G. 146, 147
strategies *see under* investment
structure of home ownership 13, 17–50
  evolution 22–33
  prices 17–22, 48–9
    see also booms; slumps
  recovery 43–6
  slump of 1990s 33–42

Sweden 21
Swennarton, M. 56
Switzerland 51
Symonds, B. 154

tax 46–7, 103–4
  reduction 174, 175, 177, 181
  relief, mortgage interest 32, 41, 54, 73, 74
  see also Inland Revenue
tenants *see* rental
tenure *see* mortgages; outright; rental
Thatcher, M. 14, 53–4, 61, 67–8, 105, 168, 184
Thomas, A. 137
Thomas, R. 7–8
Thorns, D.C. 21, 131
  investment 75, 76, 84–6, 146, 147
Thrift, N. 61
Tickell, A. 180
Torney, T. 2–5
Town and Country Building Society 6
trading down 137–8, 140–2, 143, 153–4, 155
trading up 153
  see also ladder
Troy, P.N. 55
Trustee Savings Bank 41
Tucillo, J. 21
Turnbull, P. 175
Turner, B. 21
type of house 10, 61–4
  equity distribution 108, 110, 114, 117, 118–19
  investment 88, 93–5
    strategies 151–61 *passim*

unemployment 33, 196
Ungerson, C. 56
United States 21, 51, 57, 124, 168
  future 185, 190, 193
  investment 74, 75
    strategies 146, 147–8
  wealth and economy 167–8
use value 21, 145, 149

van der Weyer, M. 183–4
Veblen, T. 61

Wales 27, 33–6, 130
  equity distribution 111, 117, 120
  structure of home ownership 27, 33–6
Walker, R. 167–8
Watt, P. 131
wealth, housing 12–13, 101–8
  BHPS data 107–8
  distribution 108–14
  and economy 13, 15–16, 167–81
    boom of 1980s 168–70, 178–80
    consumer spending 168–70, 174–8
    financial deregulation 178–80
    recession caused by housing market
      170–1
    South East 180–1

  inequality reduction 103–6
  and personal wealth 106–7
  see also inheritance
Weber, M. 65
Weiner 62
Westaway, P. 131, 133, 137, 143
White Papers 52–4
Whitehead, C. 73
Wilcox 173
Willcock, J. 42
Williams, P. 52, 55, 56, 125, 130
women 64, 104
Wood, D. 43, 184
Woolwich Building Society 6, 38
Wriglesworth, J. 110, 115–16